Workshop on Biomedical Natural Language Processing

(BioNLP 2014)

Held at ACL 2014

Baltimore, Maryland, USA
27 - 28 June 2014

ISBN: 978-1-63439-214-3

Printed from e-media with permission by:

Curran Associates, Inc.
57 Morehouse Lane
Red Hook, NY 12571

CURRAN ASSOCIATES INC. proceedings.com

Some format issues inherent in the e-media version may also appear in this print version.

Printed by Curran Associates, Inc. (2014)

Association for Computational Linguistics
209 N. Eighth Street
Stroudsburg, Pennsylvania 18360

Phone: 1-570-476-8006
Fax: 1-570-476-0860

acl@aclweb.org

Additional copies of this publication are available from:

Curran Associates, Inc.
57 Morehouse Lane
Red Hook, NY 12571 USA
Phone: 845-758-0400
Fax: 845-758-2634
Email: curran@proceedings.com
Web: www.proceedings.com

Introduction

The first day of the BioNLP 2014 workshop continues following the course set by the first ACL workshop on Natural Language Processing in the Biomedical Domain that was held in 2002: BioNLP 2014 provides a venue for exploring challenges and techniques in processing biomedical language and brings together researchers from computational linguistics and biomedical informatics. The submissions to the first day of 2014 workshop organized by SIGBioMed were traditionally very strong and continued demonstrating the considerable breadth of research in biomedical language processing. The 2014 workshop has accepted 12 full and short papers for oral presentations and 7 posters. The first day of the workshop features a keynote that expands the scope of BioNLP beyond its already remarkable breadth

Keynote BioNLP as the Pioneering field of linking text, knowledge and data

Professor Jun'ichi Tsujii, Principal Researcher at Microsoft Research Asia (MSRA), Chair of Text Mining and Scientific Director of the National Centre for Text Mining (NaCTeM) at the University of Manchester, UK

The second day of the workshop features a paper submitted to the special track on NLP approaches for assessment of clinical conditions. Kathleen C. Fraser presents the featured talk on using statistical parsing to detect agrammatic aphasia. The track organizers, Thamar Solorio and Yang Liu, serve as discussants.

The second day further features an exciting panel that brings together organizers of several shared tasks in biomedical information retrieval and natural language processing. The panel introduces the workshop participants to the long-standing and relatively new community-wide challenges in biomedical and clinical language processing. It also provides an opportunity to discuss the future of the shared tasks in this domain.

Panel Life cycles of BioCreative, BioNLP-ST, i2b2, TREC Medical tracks, and ShARe /CLEF/ SemEval

Lynette Hirschman & John Wilbur, Sophia Ananiadou, Ellen Voorhees, Ozlem Uzuner, Danielle Mowery & Sumithra Velupillai & Sameer Pradhan

The second day of the BioNLP 2014 workshop concludes with two tutorials on the fundamental resources widely used in the biomedical domain.

Tutorial 1 UMLS in biomedical text processing

Olivier Bodenreider, Branch Chief, Cognitive Science Branch, LHNCBC, NLM, NIH

Tutorial 2 Using MetaMap Alan R. Aronson, Senior Researcher, Cognitive Science Branch, LHNCBC, NLM, NIH

Acknowledgments

As always, we are profoundly grateful to the authors who chose BioNLP as venue for presenting their innovative research. The authors' willingness to share their work through BioNLP consistently makes the workshop noteworthy among the increasing numbers of available venues. We are equally indebted to the program committee members (listed elsewhere in this volume) who produced at least three thorough reviews per paper on a tight review schedule and with an admirable level of insight.

Organizers:

Kevin Bretonnel Cohen, University of Colorado School of Medicine
Dina Demner-Fushman, US National Library of Medicine
Sophia Ananiadou, University of Manchester and National Centre for Text Mining, UK
John Pestian, University of Cincinnati, Cincinnati Children's Hospital Medical Center
Jun'ichi Tsujii, Microsoft Research Asia and National Centre for Text Mining, UK

Program Committee:

Emilia Apostolova, DePaul University, USA
Eiji Aramaki, University of Tokyo, Japan
Alan Aronson, US National Library of Medicine
Sabine Bergler, Concordia University, Canada
Olivier Bodenreider, US National Library of Medicine
Kevin Cohen, University of Colorado, USA
Nigel Collier, National Institute of Informatics, Japan
Dina Demner-Fushman, US National Library of Medicine
Marcelo Fiszman, US National Library of Medicine
Filip Ginter, University of Turku, Finland
Graciela Gonzalez, Arizona State University, USA
Antonio Jimeno Yepes, NICTA, Australia
Halil Kilicoglu, US National Library of Medicine
Jin-Dong Kim, University of Tokyo, Japan
Robert Leaman, US National Library of Medicine
Yang Liu, The University of Texas at Dallas, USA
Zhiyong Lu, US National Library of Medicine
Makoto Miwa, National Centre for Text Mining, UK
Aurelie Neveol, LIMSI, France
Naoaki Okazaki, Tohoku University, Japan
Jong Park, KAIST, South Korea
Rashmi Prasad, University of Wisconsin-Milwaukee, USA
Sampo Pyysalo, National Centre for Text Mining, UK
Bastien Rance, Georges Pompidou European Hospital, France
Thomas Rindflesch, US National Library of Medicine
Kirk Roberts, US National Library of Medicine
Andrey Rzhetsky, University of Chicago, USA
Matthew Simpson, US National Library of Medicine
Thamar Solorio, The University of Alabama at Birmingham, USA
Yoshimasa Tsuruoka, University of Tokyo, Japan
Karin Verspoor, NICTA, Australia
W. John Wilbur, US National Library of Medicine

Invited Speaker:

Jun'ichi Tsujii, Microsoft Research Asia and National Centre for Text Mining, UK

Panelists

Sophia Ananiadou, University of Manchester and National Centre for Text Mining, UK
Lynette Hirschman, The MITRE Corporation , USA

Danielle Mowery, University of Pittsburgh, USA
Sameer Pradhan, Harvard Medical School, USA
Ozlem Uzuner, State University of New York, Albany, USA
Sumithra Velupillai, Stockholm University, Sweden
Ellen Voorhees, National Institute of Standards and Technology, USA
W. John Wilbur, US National Library of Medicine

Table of Contents

Conference Program

Thursday, June 26, 2014

9:00–9:10 Opening remarks

Session 1: Processing biomedical publications

9:10–9:30 *Natural Language Processing Methods for Enhancing Geographic Metadata for Phylogeography of Zoonotic Viruses*
Tasnia Tahsin, Robert Rivera, Rachel Beard, Rob Lauder, Davy Weissenbacher, Matthew Scotch, Garrick Wallstrom and Graciela Gonzalez

9:30–9:50 *Temporal Expression Recognition for Cell Cycle Phase Concepts in Biomedical Literature*
Negacy Hailu, Natalya Panteleyeva and Kevin Cohen

9:50–10:10 *Classifying Negative Findings in Biomedical Publications*
Bei Yu and Daniele Fanelli

10:10–10:30 *Automated Disease Normalization with Low Rank Approximations*
Robert Leaman and Zhiyong Lu

10:30–11:00 Coffee Break

Keynote by Junichi Tsujii

11:00–11:50 BioNLP as the Pioneering field of linking text, knowledge and data

Session 2: Processing consumer language

11:50–12:10 *Decomposing Consumer Health Questions*
Kirk Roberts, Halil Kilicoglu, Marcelo Fiszman and Dina Demner-Fushman

12:10–12:30 *Detecting Health Related Discussions in Everyday Telephone Conversations for Studying Medical Events in the Lives of Older Adults*
Golnar Sheikhshab, Izhak Shafran and Jeffrey Kaye

12:30–14:00 Lunch

Session 3: Processing clinical text and gray literature

(16:40–17:30) Poster session

Thursday, June 26, 2014 (continued)

Seeking Informativeness in Literature Based Discovery
Judita Preiss

Towards Gene Recognition from Rare and Ambiguous Abbreviations using a Filtering Approach
Matthias Hartung, Roman Klinger, Matthias Zwick and Philipp Cimiano

FFTM: A Fuzzy Feature Transformation Method for Medical Documents
Amir Karami and Aryya Gangopadhyay

Friday, June 27, 2014

Session 1: NLP approaches for assessment of clinical conditions

9:00–9:40 *Using statistical parsing to detect agrammatic aphasia*
Kathleen C. Fraser, Graeme Hirst, Jed A. Meltzer, Jennifer E. Mack and Cynthia K. Thompson

Panel: Life cycles of BioCreative, BioNLP-ST, i2b2, TREC Medical tracks, and ShARe /CLEF/ SemEval

9:40–10:05 BioCreative by Lynette Hirschman and John Wilbur

10:05–10:30 BioNLP-ST by Sophia Ananiadou and Junichi Tsujii

10:30–11:00 Coffee Break

11:00–11:25 TREC Medical tracks by Ellen Voorhees

11:25–11:50 i2b2 by Ozlem Uzuner

11:50–12:10 ShARe/CLEF/SemEval by Danielle Mowery, Sumithra Velupillai and Sameer Pradhan

12:10–12:30 Discussion

12:30–14:00 Lunch

Friday, June 27, 2014 (continued)

Tutorials

14:00–15:30 UMLS in biomedical text processing by Olivier Bodenreider

15:30–16:00 Afternoon Break

16:00–17:30 Using MetaMap by Alan R. Aronson

Natural Language Processing Methods for Enhancing Geographic Metadata for Phylogeography of Zoonotic Viruses

Tasnia Tahsin
Department of Biomedical Informatics
Arizona State University
13212 E Shea Blvd
Scottsdale, AZ 85259
ttahsin@asu.edu

Rachel Beard
Department of Biomedical Informatics
Arizona State University
13212 E Shea Blvd
Scottsdale, AZ 85259
rachel.beard@asu.edu

Robert Rivera
Department of Biomedical Informatics
Arizona State University
13212 E Shea Blvd
Scottsdale, AZ 85259
rdriver1@asu.edu

Rob Lauder
Department of Biomedical Informatics
Arizona State University
13212 E Shea Blvd
Scottsdale, AZ 85259
rlauder@asu.edu

Davy Weissenbacher
Department of Biomedical Informatics
Arizona State University
13212 E Shea Blvd
Scottsdale, AZ 85259
dweissen@asu.edu

Garrick Wallstrom
Department of Biomedical Informatics
Arizona State University
13212 E Shea Blvd
Scottsdale, AZ 85259
gwallstrom@asu.edu

Matthew Scotch
Department of Biomedical Informatics
Arizona State University
13212 E Shea Blvd
Scottsdale, AZ 85259
mscotch@asu.edu

Graciela Gonzalez
Department of Biomedical Informatics
Arizona State University
13212 E Shea Blvd
Scottsdale, AZ 85259
Graciela.gonzalez@asu.edu

Abstract

Zoonotic viruses, viruses that are transmittable between animals and humans, represent emerging or re-emerging pathogens that pose significant public health threats throughout the world. It is therefore crucial to advance current surveillance mechanisms for these viruses through outlets such as phylogeography. Phylogeographic techniques may be applied to trace the origins and geographical distribution of these viruses using sequence and location data, which are often obtained from publicly available databases such as GenBank. Despite the abundance of zoonotic viral sequence data in GenBank records, phylogeographic analysis of these viruses is greatly limited by the lack of adequate geographic metadata. Although more detailed information may often be found in the related articles referenced in these records, manual extraction of this information presents a severe bottleneck. In this work, we propose an automated system for extracting this information using Natural Language Processing (NLP) methods. In order to validate the need for such a system, we first determine the percentage of GenBank records with "insufficient" geographic metadata for seven well-studied zoonotic viruses. We then evaluate four different named entity recognition (NER) systems which may help in the automatic extraction of information from related articles that can be used to improve the GenBank geographic metadata. This includes a novel dictionary-based location tagging system that we introduce in this paper.

Proceedings of the 2014 Workshop on Biomedical Natural Language Processing (BioNLP 2014), pages 1–9,
Baltimore, Maryland USA, June 26-27 2014.

1 Introduction

Zoonotic viruses, viruses that are transmittable between animals and humans, have become increasingly prevalent in the last century leading to the rise and re-emergence of a variety of diseases (Krauss, 2003). In order to enhance currently available surveillance systems for these viruses, a better understanding of their origins and transmission patterns is required. This need has led to a greater amount of research in the field of phylogeography, the study of geographical lineages of species (Avise, 2000). Population health agencies frequently apply phylogeographic techniques to trace the evolutionary changes within viral lineages that affect their diffusion and transmission among animal and human hosts (Ciccozzi et al., 2013; Gray and Salemi, 2012; Weidmann et al., 2013). Prediction of virus migration routes enhances the chances of isolating the viral strain for vaccine production. In addition, if the source of the strain is identified, intervention methods may be applied to block the virus at the source and limit outbreaks in other areas.

Phylogeographic analysis depends on the utilization of both the sequence data and the location of collection of specific viral sequences. Researchers often use publicly available databases such as GenBank for retrieving this information. For instance, Wallace and Fitch (2008) used data from GenBank records to study the migration of the H5N1 virus in various animal hosts over Europe, Asia and Africa, and were able to identify the Guangdong province in China as the source of the outbreak. However, the extent of phylogeographic modeling is highly dependent on the specificity of available geospatial information and the lack of geographic data more specific than the state or province level may limit phylogeographic analysis and distort results. In the previous example, Wallace and Fitch (2008) had to use town-level information to identify the source of the H5N1 outbreak; without specific location data, they would not have been able to identify the Guangdong province as the source. Unfortunately, while there is an abundance of sequence data in GenBank records, many of them lack sufficient geographic metadata that would enable specific identification of the isolate's location of collection. A prior study conducted by Scotch et al. (2011) showed that the geographic information of 80% of the GenBank records associated with single or double stranded RNA viruses within tetrapod hosts is less specific than 1st level administrative boundaries (ADM1) such as state or province.

Though many of the records lack specific geographic metadata, more detailed information is often available within the journal articles referenced in them. However, manual extraction of this information is time-consuming and cumbersome and presents a severe bottleneck on phylogeographic analysis. In this work, we investigate the potential of NLP techniques to enhance the geographic data available for phylogeographic studies of zoonotic viruses using NER systems. In addition to geographic metadata and sequence information, GenBank records also contain several other forms of metadata such as host, collection date and gene for each isolate. Journal articles that are referenced in these records often mention the location of isolation for the viral sample in conjunction with related metadata (Figure 1 provides an example of such a case). Therefore, by allowing identification of location mentions along with mentions of related GenBank metadata in these articles, we believe that NER systems may help to accurately link each GenBank record to its corresponding location of isolation and distinguish it from other location mentions.

Previously Scotch et al. (2011) evaluated the performance of BANNER (Leaman and Gonzalez, 2008) and the Stanford NER tool (Finkel et al., 2005) for automated identification of gene and location mentions respectively, in 10 full-text PubMed articles, each related to a specific GenBank record. They were both found to achieve f-scores of less than 0.45, thereby establishing the need for NER systems with better performance and/or a larger test corpus (Scotch et al, 2011). In this study, we start by evaluating the state of geographic insufficiency for zoonotic viruses in GenBank records using a new automated approach. Next, we further expand upon the work done by Scotch et al. (2011) by building our own dictionary-based location-tagging system and evaluating its performance on a larger corpus corresponding to over 8,500 GenBank records for zoonotic viruses. In addition, we also evaluate the performance of three other state-of-the-art NER tools for tagging gene, date and species mentions in this corpus. We believe that identification of these entities will be useful for the future development of a system for extracting the location of collection of viral isolates from articles related to their respective GenBank records.

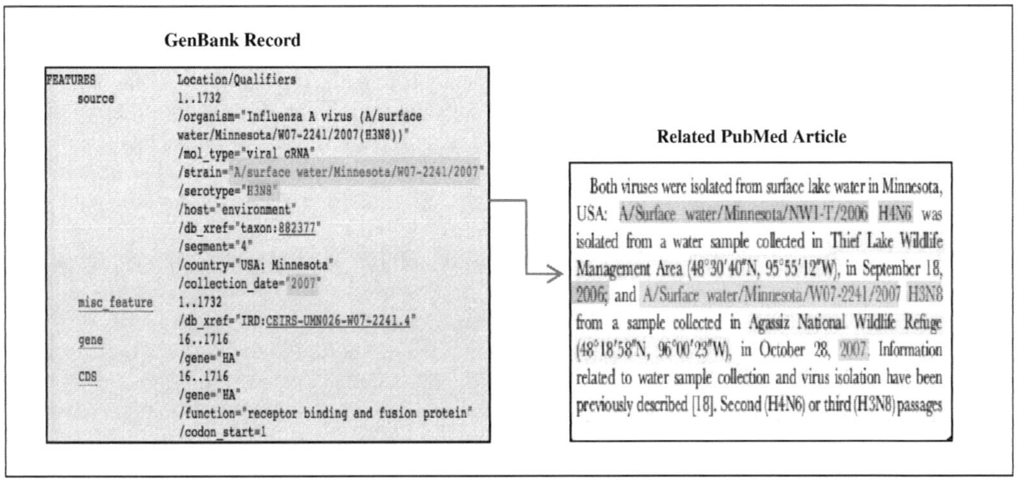

Figure 1. Example of how the date, gene, and strain metadata within a GenBank record may be used to differentiate between two potential locations in a related article

2 Methods

The process undertaken to complete this study can be divided into three distinct stages: selection of the zoonotic viruses and extraction of relevant GenBank data related to each virus, computation of "sufficiency" statistics on the extracted data, and development/evaluation of NER systems for tagging location, gene, date and species mentions in full-text PubMed Central articles. A detailed description of each phase is given below.

2.1 Virus Selection and GenBank Data Extraction

The domain of this study has been limited to zoonotic viruses that are most consistently documented and tracked by public health, agriculture and wildlife state departments within the United States. These viruses include influenza, rabies, hantavirus, western equine encephalitis (WEE), eastern equine encephalitis (EEE), St. Louis encephalitis (SLE), and West Nile virus (WNV). The Entrez Programming Utilities (E-Utilities) was used to download the following fields from 59,595 GenBank records associated with these viruses: GenBank Accession ID, PubMed Central ID, Strain name, Collection date and Country. These records were the result of a query performed to retrieve all accession numbers related to the selected viruses which had at least one reference to a PubMed Central article. The results from the query was retrieved on August 22nd, 2013.

2.2 Sufficiency Analysis

Database Integration: The data extracted from Genbank was used to compute the percentage of GenBank records that had insufficient geographic information for each of the selected viruses. In order to perform this computation, we used data from the ISO 3166-1 alpha-2 [1] table and the GeoNames database. The ISO 3166-1 alpha-2 is the International Standard for representing country names using two-letter codes. The GeoNames[2] database contains a variety of geospatial data for over 10 million locations on earth, including the ISO 3166-1 alpha-2 code for the country of each location and a feature code that can be used to determine the administrative level of each location. To allow for efficient querying, we downloaded the main GeoNames table and the ISO alpha-2 country codes table from their respective websites and stored them in a local SQL database. Prior to adding the ISO data to the database, some commonly used country names and their corresponding country codes were added to the table since it only included a single title for each country. For example, the ISO table included the country name "United States" but not alternate names such as "USA", "United States of America", or "US". Using the created database in conjunction with a parser written in Java, we were able to retrieve most

[1] Iso.org. [Internet]. Genève. c2013. Available from http://www.iso.org/iso/home/standards/country_codes.htm

[2] Geonames.org. [Internet]. Egypt. c2013. [updated 2013 Apr 30] Available from http://www.geonamesorg/EG/administrative-division-egypt.html

of the geographic information present within the records and classify each of them as sufficient or insufficient.

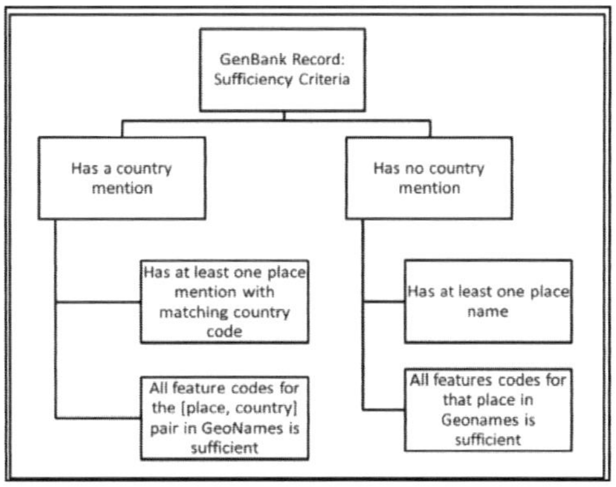

Figure 2. Sufficiency Criteria

Sufficiency Criteria: For the purpose of this project, we considered any geographical boundary more specific than ADM1 to be "sufficient". Based on this criterion, a feature code in GeoNames was categorized as sufficient only if it was absent from the following list of feature codes: ADM1, ADM1H, ADMD, ADMDH, PCL, PCLD, PCLF, PCLH, PCLI and PCLS. Evaluation of the geographical sufficiency of a GenBank record was dependent upon whether the record included a country name. A GenBank record with a country mention was called sufficient if the geographic information extracted from that record included another place mention whose feature code fell within the class of sufficient feature codes and whose ISO country code matched that of the retrieved country. For instance, a GenBank record with the geographic metadata "Orange County, United States" will be called sufficient since the place "Orange County" has a sufficient feature code of "ADM2" and a country code of "US" which matches the country code of the retrieved country, "United States". Place mentions with matching country codes often had several different feature codes in GeoNames. Such places were only called sufficient if all feature codes corresponding to the given pair of place name and country code were classified as sufficient. In cases where the GenBank record had no country mention, the record was called sufficient only if all matching GeoNames entries for any of the places mentioned in it had sufficient feature codes. The sufficiency criteria were designed to ensure that a geographic location is only called sufficient if its administrative level was found to be more specific than ADM1 without any form of ambiguity. Figure 3 illustrates the pathways of geographical sufficiency for GenBank records in a diagram.

Sufficiency Computation: In order to obtain the geographic information for each Genbank record, we used a Java parser which automatically extracted data from the "country" field of each record. Since the "country" field typically contained multiple place mentions divided by a set of delimiters consisting of comma, colon and hyphen, we first split this field using these delimiters. We then checked each string obtained through this process against the ISO country code table to determine whether it was a potential country name for the record's location. If the query returned no results, then the locally stored GeoNames table was searched and for each match found, the corresponding ISO country code and feature code were extracted. Figure 4 shows a diagram of this process.

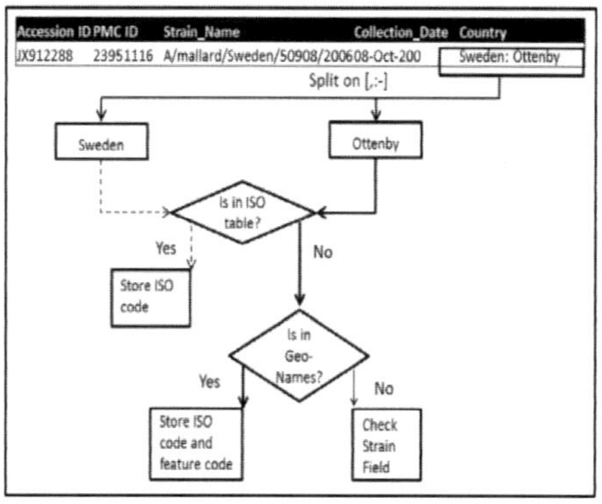

Figure 3. Sufficiency Calculation Example

In cases where no sufficient location data was found from the "country" field of a GenBank record, the Java parser searched through its "strain" field. This was done because some viral strains such as influenza include their location of origin integrated into their names. For example, the influenza strain "A/duck/Alberta/35/76" indicates that the geographic origin of the strain is Alberta. The different sections of a strain field are separated by either forward slash, parenthesis, comma, colon, hyphen or underscore and so we used a set of delimiters consisting of these characters to split this field. Each string thus retrieved was queried as before on the ISO country code table and the GeoNames table. GeoNames often returned matches for strings like 'raccoons' and 'chicken' which were actually meant to be names of host species within the "strain" field, and so a list of

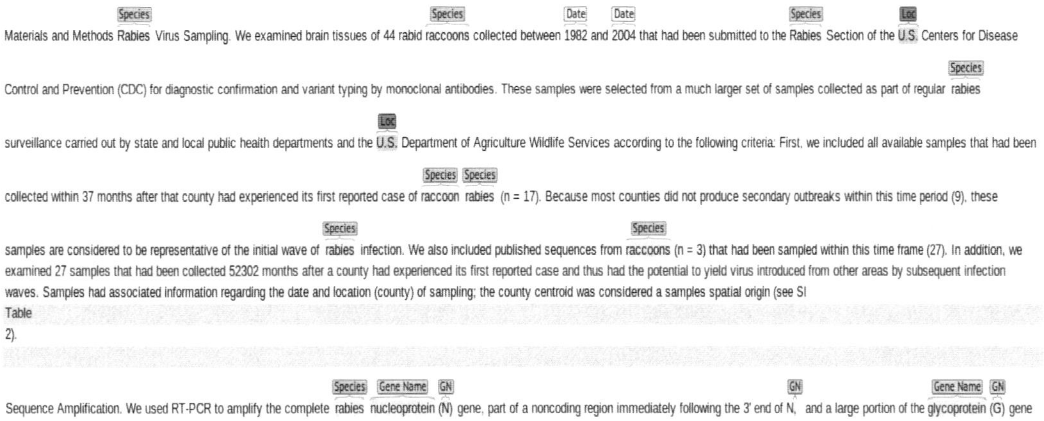

Figure 4. Example of annotation including all four entities

some of the most frequently seen host name mentions in these records was manually created and filtered out before querying GeoNames.

Some of the place mentions contained very specific location information which resulted in GeoNames not finding a match for them. A list was created for strings like 'north', 'south-east', 'governorate' etc. which when removed from a place mention may produce a match. In cases of potential place mentions which contained any one of these strings and for which GeoNames returned no matching result, a second query was performed after removal of the string.

Evaluation of Sufficiency Computation: We manually annotated 10% of all influenza records in GenBank which reference at least one PubMed Central article as sufficient or insufficient based on our sufficiency criteria (5731 records). We then ran our program on these records and compared system results with annotated results.

2.3 Development/Evaluation of NER systems

Creation of Gold Standard Corpus: We created a gold standard corpus consisting of twenty-seven manually-annotated full-text PubMed Central articles in order to evaluate the performance of NER systems for tagging location, gene, species and date mentions in text. The articles corresponded to over 8,500 GenBank records and were randomly sampled using the subset of extracted GenBank records which contained a link to PubMed Central articles and had insufficient geographic metadata.

Three annotators tagged the following four entities in each article using the freely available annotation tool, BRAT (Stenetorp et al., 2012): gene names, locations, dates and species. Figure 4 provides an example of the manual annotation in BRAT. We annotated all mentions of each entity type, not only those relevant to zoonotic viruses, in order to evaluate system performance. A total of over 19,000 entities were annotated within this corpus. The number of tokens annotated was about 24,000. A set of annotation guidelines was created for this process (available upon request). Before creating the guidelines, each annotator individually annotated six common articles and compared and discussed their results to devise a reasonable set of rules for annotating each entity. After discussion, the annotators re-annotated the common articles based on the guidelines and divided the remaining articles amongst themselves. The inter-annotator agreement was calculated for each pair of annotators. The annotated corpus will be made available at diego.asu.edu/downloads.

Development of Automated Location Tagger: We developed a dictionary-based NER system using the GeoNames database for automated identification of location mentions in text. The dictionary used by this system, which we will hereby refer to as GeoNamer, was created by retrieving distinct place names from the GeoNames table and filtering out commonly used words from the retrieved set. Words filtered out include stop words such as 'are' and 'the', generic place names such as 'cave' and 'hill', numbers like 'one' and 'two', domain specific words such as 'biology' and 'DNA', most commonly used surnames like 'Garcia', commonly used animal names such as 'chicken' and 'fox' and other miscellaneous words such as 'central'. This was a crucial step since the GeoNames database contains a wide array of commonly used English words which may cause a large volume of false positives if not removed. The final dictionary consists of 5,396,503 entries. In order to recognize place mentions in a given set of text files, GeoNamer first builds a Lucene index on the contents of the files. It then constructs a phrase query for every entry in the Geonames dictionary and runs each query on the Lucene index. The document id, query text, start offset and end offset for every match found is written to an output file. We chose this approach because of its simplicity and efficiency.

Evaluation of NER Systems: Four different NER systems for identifying species, gene, date and location mentions in text were evaluated using the created gold standard. The evaluated systems include LINNEAUS (Gerner et al., 2010), BANNER, Stanford SUTime (Chang and Manning, 2012) and GeoNamer. LINNEAUS, BANNER and Stanford SUTime are widely-used, state-of-the-art open source NER systems for recognition of species, gene and temporal expressions respectively. GeoNamer is the system we developed in this work for the purpose of tagging locations, as described earlier.

3 Results

3.1 Sufficiency Analysis

The system for classifying records as sufficient or insufficient was found to have an accuracy of 72% as compared to manual annotation. 98% of the errors was due to insufficient records being called sufficient. The results of the sufficiency analysis are given in Table 1. 64% of all GenBank records extracted for this project contained insufficient geographic information. Amongst the seven studied viruses, WEE had the highest and EEE had the lowest percentage of insufficient records.

Virus Type	Number of Entries	% Insufficient
WEE	67	90
Rabies	4450	85
WNV	1084	79
SLE	141	74
Hanta	1745	66
Influenza	51734	62
EEE	374	51
All	59595	64

Table 1. Percentage of GenBank records with insufficient geographic information for each zoonotic virus studied in this project

3.2 Gold Standard Corpus

The results for the comparison of the annotations performed by our three annotators on 6 common papers can be found in Table 2. We used the F-score between each pair of annotators as a measure of inter-rater agreement and had over 90% agreement with overlap matching and over 86% agreement with exact matching in all cases. The final gold standard corpus contained approximately 19,000 entities corresponding to approximately 24,000 tokens.

Entity	F-score (*A,B*) (Exact; Overlap)	F-score (*A,C*) (Exact; Overlap)	F-score (*B,C*) (Exact; Overlap)
Date	.975; .978	.979; .987	.962; .973
Gene	.914; .926	.913; .932	.911; .954
Location	.945; .961	.907; .931	.914; .935
Species	.909; .956	.874; .940	.915; .959
Virus	.952; .958	.947; .966	.947; .955
Mean	**.939; .956**	**.924; .951**	**.930; .955**

Table 2. Frequency of Annotated Entities for 6 common annotated papers

3.3 Performance Analysis of NER Systems

The performance metrics for the NER systems at tagging the desired entities in the test set are listed in Table 3. The highest performance was achieved by Stanford SUTime for date tagging. Tagging of genes had the lowest performance.

Entity	Precision (Exact; Overlap)	Recall (Exact; Overlap)	F-score (Exact; Overlap)
BANNER	0.070; 0.239	0.114; 0.395	0.087; 0.297
GeoNamer	0.452; 0.626	0.658; 0.783	0.536; 0.696
LINNEAUS	0.853; 0.962	0.563; 0.658	0.678; 0.781
Stanford SUTime	0.800; 0853	0.681; 0.727	0.736; 0.785

Table 3. Performance Statistics of NER

4 Discussion

Based on our analysis, at least half of the GenBank records for each of the studied zoonotic viruses lack sufficient geographic information, and the proportion of insufficient records can be as high as 90%. Our automated system for classifying records as insufficient or sufficient was found to have an accuracy of 72% with 98% of the errors being a result of insufficient records being called sufficient. Therefore, our computed estimate of insufficiency is very likely to be an underestimation of the actual problem. The virus with the highest level of sufficiency, EEE, had a large number of records with county level information in the "country" field. However, the insufficient records for this virus typically contained no place mention, not even at the country level. A key reason for our calculated percentage of sufficient GenBank records being higher for these seven viruses than what has been previously computed by Scotch et al. (2011) was the inclusion of the "strain" field. The "strain" field often contained specific location information which, when combined with place mentions present within the "country" field, made the record geographically sufficient. The virus for which the inclusion of "strain" field had the greatest impact on boosting the sufficiency percentage was influenza. Most of the GenBank records associated with this virus had structured "strain" fields from which the parser could easily separate place mentions using GeoNames.

Although the sufficiency classifications produced by our system were correct most of the time, there were a few cases where a record got incorrectly labeled as insufficient even when it contained detailed geographic information. This typically happened because GeoNames failed to return matching results for these places. For instance, the country field "India: Majiara,WB" was not found to be sufficient even though Majiara is a city in India because GeoNames has no entry for it. In some cases the lack of matching result was due to spelling variations of the place name. For instance the country field "Indonesia: Yogjakarta" was called insufficient since "Yogjakarta" is spelled as "Yogyakarta" in GeoNames. Sometimes the database simply did not contain the exact string present in the GenBank record. For instance, it does not have any entry for the place "south Kalimantan" but it contains the place name "kalimantan". The number of sufficient records which were called insufficient by our system due to inexact matching were greatly mitigated by removing strings such as "south" from the place mention, as described in the "Methods" section.

Most of the NER systems performed significantly better with overlap measures than with exact-match measures. This is because our annotation guidelines typically involved tagging the longest possible match for each entity and the automated systems frequently missed portions of each annotation. Stanford SUTime had the best overlap f-measure of 0.785, closely followed by LINNEAUS with an overlap f-measure of 0.781. Although Stanford SUTime was fairly effective at finding date mentions in text, it tagged all four-digit-numbers such as "1012" and "2339" as years, leading to a number of false positives. The poor recall of LINNEAUS was mostly caused because the dictionary used by LINNEAUS tagged only species mentions in text while we tagged genus and family mentions as well. It also missed a lot of commonly used animal names such as monkey, bat, badger and wolf. GeoNamer was the third best performer with the highest recall but second lowest precision. This is because the GeoNames dictionary contains an extensively large list of location names, many of which are commonly used words such as "central". Even though we filtered out a vast majority of these words, it still produced false positives such as "wizard". However, its performance was considerably better than that of the Stanford location tagger used by Scotch et al. (2011) which was found to have a recall, precision and f-score of 0.26, 0.81 and 0.39 respectively. The improved performance was achieved because of the higher recall of our system. The GeoNames dictionary provides an extensive coverage of all location mentions in the world and the Stanford NER system, which is a CRF classifier trained on a different dataset, was not able to recognize many of the place mentions present in full-text PMC articles related to GenBank records.

BANNER showed the poorest performance amongst all the entity taggers evaluated in this paper. In fact, the f-score we achieved for BANNER in this study was much lower that its past f-score of 0.42 within the domain of articles related to GenBank records for viral isolates (Scotch et al., 2011). As mentioned by Scotch et al. (2011), a key reason for BANNER's poor performance in this domain is the difference between the data set used to train the BANNER model and the annotation corpus used to test this system. The version of BANNER used in these two studies was trained on the training set for the BioCreative 2 Gene

Mention task, which comprised of 15,000 sentences from PubMed abstracts. These abstracts often contained the full names for gene and protein mentions while the full-text articles we used mostly contained the abbreviated forms of gene names, which BANNER tended to miss. The articles also contained abbreviated forms of several entities such as viral strain name (e.g. H1N1) and species name (e.g. VEEV) which look similar to abbreviated gene names. Therefore, BANNER often misclassified these entities as gene mentions. A possible reason for BANNER having a much lower performance in this study than in the previous study conducted by Scotch et al (2011) is the presence of a large number of tables in the journal articles we selected. BANNER is a machine learning system based on conditional random fields which uses orthographic, morphological and shallow syntax features extracted from sentences to identify gene mentions in text. Such features do not help greatly for extraction from tables. Therefore, BANNER was often not able to identify the gene mentions in the tables present within our corpus, thereby producing false negatives. Moreover, it tagged several entries within the table as a single gene name, thereby producing false positives as well. This reduced both the recall and precision of BANNER.

Although this study explores the problem of insufficient geographic information in GenBank more thoroughly than past studies, the number of papers annotated as the gold standard is still limited. Thus, the performance of the taggers reported can be construed as a preliminary estimate at best. The set of taggers and their performance seem to be adequate for a large-scale application, with the exception of BANNER. However, we did not make any changes to the BANNER system (specifically, re-training) since changes to it are not possible until sufficient data is annotated for retraining.

5 Conclusions and Future Work

It can be concluded that the majority of GenBank records for zoonotic viruses do not contain sufficient geographic information concerning their origin. In order to enable phylogeographic analysis of these viruses and thereby monitor their spread, it is essential to develop an efficient mechanism for extracting this information from published articles. Automated NER systems may help accelerate this process significantly. Our results indicate that the NER systems LINNEAUS, Stanford SUTime and GeoNamer produce satisfactory performance in this domain and thus can be used in the future for linking GenBank records with their corresponding geographic information. However, the current version of BANNER is not well-suited for this task. We will need to train BANNER specifically for this purpose before incorporating it within our system.

We are currently altering the component of our program which classifies records as sufficient or insufficient in order to reduce the number of errors due to insufficient records being called sufficient. We are also manually looking through GenBank records for zoonotic viruses with insufficient geographic metadata and linking them to the location mentions in related articles which we deem to be the most likely location of collection for the given viral isolate. The resulting annotated corpus will be used to train and evaluate an automated system for populating GenBank geographic metadata. We have already covered all GenBank records related to Encephalitis viruses and close to 10% of all records related to Influenza which are linked to PubMed Central articles. The annotation process has revealed that a large proportion of the information allowing linkage of GenBank records to geographic metadata is often present in tables within the articles in addition to textual sentences. Therefore, we have developed a Python parser for automatically linking GenBank records to location mentions using tables from the HTML version of the PubMed Central articles. Future work will include further expansion of this annotation corpus and the development of an integrated system for enhancing GenBank geographic metadata for phylogeographic analysis of zoonotic viruses.

Acknowledgement

Research reported in this publication was supported by the NIAID of the NIH under Award Number R56AI102559 to MS and GG. The content is solely the responsibility of the authors and does not necessarily represent the official views of the National Institutes of Health

References

Avise, John C. (2000). Phylogeography : the history and formation of species Cambridge, Mass.: Harvard University Press.

Chang, Angel X., and Christopher Manning. "SUTime: A library for recognizing and normalizing time expressions." LREC. 2012.

Ciccozzi M, et al. Epidemiological history and phylogeography of West Nile virus lineage 2. Infection, Genetics and Evolution. 2013:17;46-50.

Finkel JR, Grenager T, Manning C. Incorporating non-local information into information extraction systems by Gibbs sampling. In: Proceedings of the 43rd annual meeting of the association for computational linguistics (ACL 2005); 2005. p. 363–70.

Gerner M, Nenadic G, and Bergman CM. LINNAEUS: A species name identification system for biomedical literature. BMC Bioinformatics. 2010;11(85).

Gray RR, and Salemi M. Integrative molecular phylogeography in the context of infectious diseases on the human-animal interface. Parasitology-Cambridge. 2012;139:1939-1951

Krauss, H. (2003). Zoonoses: infectious diseases transmissible from animals to humans (3rd ed.). Washington, D.C.: ASM Press.

Leaman R and Gonzalez G. BANNER: An executable survey of advances in biomedical named entity recognition. Pacific Symposium on Biocomputing. 2008;13:652-663.

Scotch, Matthew, et al. Enhancing phylogeography by improving geographical information from GenBank. Journal of biomedical informatics. 2011;44:S44-S47.

Stenetorp P, et al. BRAT: A Web-based Tool for NLP-Assisted Text Annotation. EACL '12 Proceedings

Wallace, R.G. and W.M. Fitch, Influenza A H5N1 immigration is filtered out at some international borders. PLoS One, 2008. 3(2): p. e1697.

Weidmann M, et al. Molecular phylogeography of tick-borne encephalitis virus in Central Europe. Journal of General Virology. 2013;94:2129-2139.

Temporal Expression Recognition for Cell Cycle Phase Concepts in Biomedical Literature

Negacy D. Hailu, Natalya Panteleyeva and **K. Bretonnel Cohen**
Computational Bioscience Program, University of Colorado Denver
School of Medicine
negacy.hailu@ucdenver.edu, natalya.panteleyeva@ucdenver.edu,
kevin.cohen@gmail.com

Abstract

In this paper, we present a system for recognizing temporal expressions related to cell cycle phase (CCP) concepts in biomedical literature. We identified 11 classes of cell cycle related temporal expressions, for which we made extensions to TIMEX3, arranging them in an ontology derived from the Gene Ontology. We annotated 310 abstracts from PubMed. Annotation guidelines were developed, consistent with existing time-related annotation guidelines for TimeML. Two annotators participated in the annotation. We achieved an inter-annotator agreement of 0.79 for an exact span match and 0.82 for relaxed constraints. Our approach is a hybrid of machine learning to recognize temporal expressions and a rule-based approach to map them to the ontology. We trained a named entity recognizer using Conditional Random Fields (CRF) models. An off-the-shelf implementation of the linear chain CRF model was used. We obtained an F-score of 0.77 for temporal expression recognition. We achieved 0.79 macro-averagee F-score and 0.78 micro-averaged F-score for mapping to the ontology.

1 Introduction

Storing and processing temporal data in biomedical informatics is important, but challenging (Zhou and Hripcsak, 2007; Augusto, 2005). Biomedical data is often intrinsically associated with time. For example, data from electronic medical records are on a clinical timeline (Zhou and Hripcsak, 2007) which links all information on the progress of a patient's status. Temporal reasoning remains a challenge for medical information systems (Combi and Shahar, 1997). Conventionally, dictionaries define time as "The continuous passage of existence in which events pass from a state of potentiality in the future, through the present, to a state of finality in the past" (Editorial Staff, undated). This traditional linear concept of temporality does not adequately capture the cyclical nature of some important biological processes, such as the cell cycle and circadian rhythms. In this paper, we describe a system for the recognition of temporal expressions related to cell cycle phases in biomedical literature. The cell cycle is a phenomenon that a cell goes through during its growth and replication. Its stages are depicted in Figure 1. We treat each phase as a distinct time component and we aim at recognizing expressions that describe them in biomedical literature, then mapping them to an ontology of cell cycle phases and transitions. Specifically, we are interested in recognizing expressions that contain one or more of the concepts shown in Table 1, where the Gene Ontology is taken as definitional of concepts related to phases of the cell cycle.

Recognition of cell cycle phase concepts from text is a non-trivial problem. Some of the ways that they can be mentioned in text, such as *interphase, anaphase,* and *prophase* are relatively unambiguous and can be recognized and mapped to an ontology using regular expressions. However, as is often the case both in general language and in biomedical language, many of the ways in which they can be mentioned are highly ambiguous. For example, *M*, which stands for mitosis, is often a unit of measurement, as in ... *removal of histone HI with 0,6 M NaCl.* (PMID: 6183061) *M* could also be an abbreviation of an author's first name, as in ... *Suzuki S, Nakata M.* (PMID: 23844291) *S*, which refers to S-phase or synthesis phase, could also stand for an author's first name, as well as a protein name, as in ... *Protein S acts as a cofactor for tissue factor pathway inhibitor.* (PMID: 23841464). In addition, the word *synthesis* is in it-

Proceedings of the 2014 Workshop on Biomedical Natural Language Processing (BioNLP 2014), pages 10–18,
Baltimore, Maryland USA, June 26-27 2014.

self ambiguous, even in the context of other mentions of cell cycle phases. In the following examples, it refers to something other than a cell cycle phase:

- *...histone synthesis by lymphocytes in G0 and G1.* (PMID: 6849885)
- *...metaphase-anaphase transition, as a result of fertilization, activation or protein synthesis inhibition.* (PMID: 9552372)

We treated recognition of temporal expressions from literature as a named entity recognition (NER) problem. Many approaches to named entity recognition are based on machine learning techniques. Nadeau and Sekine report that although semi-supervised learning algorithms have been employed in NER challenges, most systems that perform well are built based on supervised learning techniques (Nadeau and Sekine, 2007). Based on this survey report, we used Conditional Random Fields (CRFs) for the recognition phase of our approach. The details of our methods are described in section 4.2.

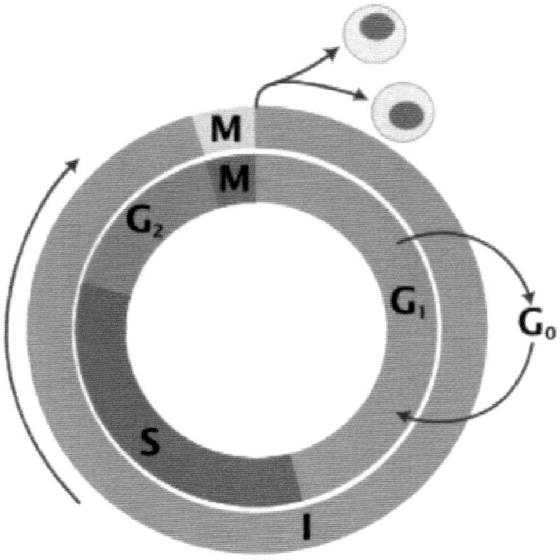

Figure 1: Schematic of the cell cycle. Outer ring: I = Interphase, M = Mitosis; inner ring: M = Mitosis, G1 = Gap 1, G2 = Gap 2, S = Synthesis; not in ring: G0 = Gap 0/Resting [Wikipedia].

2 Motivation

A vast collection of biomedical literature in PubMed/MEDLINE and other biomedical journal repositories is estimated to grow exponentially (Hunter and Cohen, 2006), as shown in Figure 2. Searching for papers specific to a researcher's interest in any domain is difficult. PubMed/MEDLINE allows search using keywords, but until recently did not rank results by document relevance. General-purpose search engines such as Google and Bing rank their results, but are not well-suited for search of specialized information related to genes and small molecules. Building a specialized search engine exclusively to search biomedical literature using genes and small molecules as keywords could be very useful, for instance, for cancer researchers.

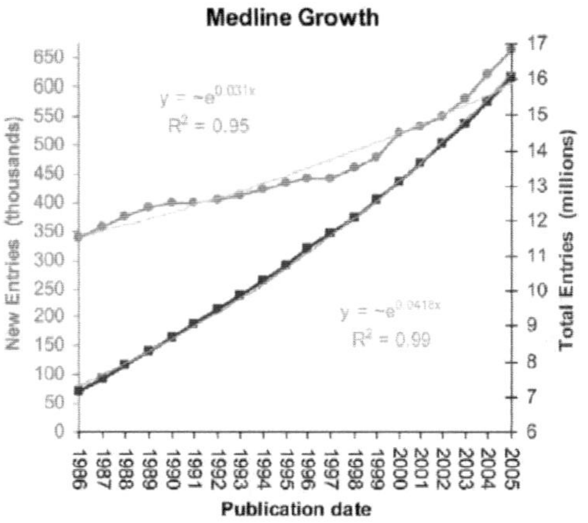

Figure 2: Publication growth rate at Medline (Hunter and Cohen, 2006)

Our long term goal is to build a specialized search engine specific to cancer research. The system will retrieve articles from PubMed/MEDLINE and rank them according to their relevance. The system will utilize gene, protein, and small molecule names as keywords in document search. We are also interested in identifying the phase(s) of the cell cycle during which the gene is expressed. After detecting the active phase(s) of a gene or gene product, the system will link relevant documents to this gene from PubMed/MEDLINE. In this paper we present our first step towards that goal, which is extraction of temporal expressions from biomedical literature. Temporal expressions will be used to identify active phases of genes or gene products.

3 Related Work

Automatic recognition of events and temporal expressions from text has attracted researchers from areas such as computer science and linguistics.

Concept	ID	Activities in each phase	Synonyms
Interphase	GO:0051325	The cell readies itself for meiosis or mitosis and the replication of its DNA occurs.	karyostasis
G0 phase	GO:0044838	Cells enter in response to cues from the cell's environment.	quiescence
G1 phase	GO:0051318	Gap phase	-
S phase	GO:0051320	DNA synthesis takes place.	S-phase, synthesis
G2 phase	GO:0051319	Gap phase	-
Mitosis	GO:0007067	The nucleus of a eukaryotic cell divides	-
Prophase	GO:0051324	Chromosomes condense and the two daughter centrioles and their asters migrate toward the poles of the cell.	-
Metaphase	GO:0051323	Chromosomes become aligned on the equatorial plate of the cell.	-
Anaphase	GO:0051322	The chromosomes separate and migrate towards the poles of the spindle.	-
Telophase	GO:0051326	The chromosomes arrive at the poles of the cell and the division of the cytoplasm starts.	-

Table 1: Cell cycle phase concepts. Definitions from the Gene Ontology.

The results have contributed to the development of diverse natural language processing applications, such as information extraction, information retrieval, question-answering systems, text summarization, etc. TimeML: Robust Specification of Event and Temporal Expressions in Text (Pustejovsky et al., 2003) is a specification language for annotation of events and temporal expressions in human language. TimeML addresses specification issues like time stamping, order of events, reasoning about events, and time expressions.

TempEval is one of the shared challenges included in SemEval (Agirre et al., 2009) as of 2007. It aims at advancing research on processing temporal information. Primarily it focuses on three tasks: event extraction and classification, temporal expression extraction and normalization, and temporal relation extraction (UzZaman et al., 2013). However, this ongoing work on temporal evaluation is based on language data collected from the news. In the clinical domain, (Styler IV et al., Undated; Palmer and Pustejovsky, 2012; Albright et al., 2013) describe the THYME annotation project. The scope and language of temporality related to the cell cycle is different from that of both TempEval and the clinical domain, and supports (and demands) different types of reasoning, specifically related to cyclical time.

Cyclical phenomena are ubiquitous in cancer development and progression. The connection between the cell cycle and cancer is well known (Vermeulen et al., 2003; Kastan and Bartek, 2004; Malumbres and Barbacid, 2009), and the fact that the cell cycle is the main target for cancer regulation, deregulation, and therapy is well established (Vermeulen et al., 2003; Kastan and Bartek, 2004; Malumbres and Barbacid, 2009). Circadian rhythms, rounds of chemotherapy, remissions, and re-occurrences all have a cyclic nature.Circadian rhythms have been investigated in the study of cancer treatment (Sahar and Sassone-Corsi, 2009; Ortiz-Tudela et al., 2013; Lengyel et al., 2009; Kelleher et al., 2014).

From the perspective of cancer research, identifying cell cycle concepts in the literature is crucial to being able to retrieve and explore information related to cyclical biological processes like the cell life cycle. From the natural language processing perspective, the novelty of this work consists in modeling cyclical time. To our knowledge, temporal event recognition grounded in a cyclical model of time has not been previously proposed.

4 Methodology

4.1 Materials

We built a corpus of 360 abstracts, consisting of 70,570 words. The concepts are presented in Table 1. We balanced our corpus by collecting articles from the PubMed/MEDLINE database using the concepts individually as keywords. We used the PubMed/MEDLINE[1] and BioMedLib search engines[2], two keyword-based search engines built on top of MEDLINE, for this purpose. The following keywords were used to collect the abstracts from PubMed and BioMedLib:

- interphase, G0, G0 phase, G1, G1 phase, synthesis, S phase, G2, G2 phase
- Mitosis, M phase, prophase, metaphase, anaphase, telophase
- checkpoint

The annotation guidelines addressed the following issues:

- The goal of the project: the goal of the annotation project was to develop a highly annotated corpus specific to CCP concepts, which will be used for automatic recognition and classification.
- Specification of each tag: this is shown in Figure 3.
- Tool used to annotate the project: We used Knowtator (Ogren, 2006), a text annotation tool built on top of the Protégé knowledge representation system.

Modeling the phenomenon was the first step in understanding the annotation process (Pustejovsky and Stubbs, 2012). We modeled our corpus as a triple, Model = <T, R, I>, as shown below:

- Model = <T, R, I> where T = terms, R = relation between the terms, and I = interaction
- T = {Named Entity, time expression, not time expression}
- R = {Named_Entity ::= TIMEXCCP | not TIMEXCCP}
- I = {TIMEXCCP = list of concepts from Table 1 or checkpoints. Examples of checkpoints are G1/G2 phase, S/G2 phase, etc.

TimeML is a specification for annotating human language in text (Pustejovsky et al., 2003). TIMEX3 is defined in TimeML as a tag for capturing dates, times, durations, and sets of dates and times. In our work we extended TIMEX3. We employ a single tag set called TIMEXCCP, where the naming is intended to be consistent with existing time-related tag sets. Figure 3 shows the attributes and functions of the tag TIMEXCCP, as well as examples of usage.

Attribute	Function	Example
Value	Interphase, G0, G1, S, G2, M, prophase, metaphase, anaphase, telophase, checkpoint.	The <TIMEXCCP value ="G1 checkpoint"> G1 checkpoint </TIMEXCCP> control mechanism ensures that everything is ready ...
Modifier	Handles temporal modifiers such as early, mid, late ...	Apoptosis induction and <TIMEXCCP modifier = "early" value = "G2/M"> early G2/M </TIMEXCCP>arrest of ...
Set	A boolean value if the time expression is a set or not.	
Comments	Comments by the annotators.	

Figure 3: Attributes and functions of the TIMEX-CCP tag.

Two annotators with training in the domain performed the annotation. Inter-annotator agreement was calculated as F-measure, following (Hripcsak and Rothschild, 2005). Inter-annotator agreement was 0.79 for an exact-span match and 0.82 for relaxed matching. The constraints, which are values of the attributes, were not considered while computing IAA for the latter case.

The annotation effort developed through several iterations, applying the annotation development cycle introduced by Pustejovsky and Stubbs (Pustejovsky and Stubbs, 2012). This methodology is depicted in Figure 4. It is called the MATTER cycle, which stands for Model, Annotation, Train, Test, Evaluation, Revise. The advantage of this methodology is that it allows us to discover hidden specifications and refine them during the MATTER cycle.

4.2 Methods

We are particularly interested in recognizing and classifying temporal expressions in the literature. For example, in the following sentence, taken from Wikipedia, the recognition task is to recognize the blue boxes as shown below and classify them. The mapping task is to categorize the recognized temporal expressions into the concepts shown in Table 1.

```
"Microhomology-mediated end
joining (MMEJ) uses a Ku
protein and DNA-PK independent
repair mechanism, and
```

[1] http://www.ncbi.nlm.nih.gov/pubmed/
[2] http://bmlsearch.com/

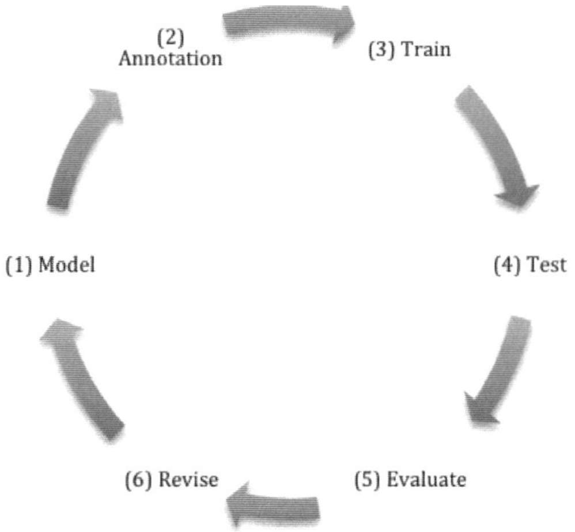

Figure 4: The MATTER cycle (Pustejovsky and Stubbs, 2012)

```
repair occurs during the
S phase of the cell cycle,
as oposed to the G0/G1 and
early S phases in NHEJ and
late S to G2 phases in HR."
```

... the [S] phase of the cell cycle, as opposed to the [G0/G1] and [early S] phases in NHEJ and [late S to G2] phase in HR.

In this example, there are four temporal expressions: *S*, *G0/G1*, *early S*, and *late S to G2*. The expression "S" is of the type S-phase or synthesis phase according to the conceptual ontology in Table 1. The expression "G0/G1" can be classified as G0 and G1. Similarly, the expression "late S to G2" can be of type S and G2.

Our approach is a hybrid of machine learning and rule-based techniques. The machine learning technique, which we refer to as the first layer, is applied for temporal expression recognition. In this layer, CRFs are trained to learn to recognize the expressions from the list of features which is shown below.

1. Word-level features:
   - Is the word in uppercase?
   - Is the first character of the word in uppercase?
   - Words themselves are also treated as features.
   - Length of the word.
2. Punctuation-related features:
   - Does the word contain at least one of the most common punctuation marks?
3. Digit-related features:
   - Is the word a digit?
   - Does the word contain a digit?
4. Does the word contain either of the following: *phase, arrest, entry*? These words typically come before or after the cell cycle concepts. For example, *early mitosis, G0 phase*.
5. Part-of-Speech tagging: Window size of 2 before and after the word.
6. Presence of concept modifiers before the word. Modifiers include: *early, mid, late, early-mid*.

Conditional Random Fields (CRFs) are one of the probabilistic graphical model sequence tagging techniques. They are understood as a sequential version of Maximum Entropy Models (Klinger and Tomanek, 2007). One advantage of CRFs over other probabilistic models like Hidden Markov Models and Maximum Entropy Models for complex systems is their support for features interacting with one another. The linear chain CRF representation is shown in Figure 5.

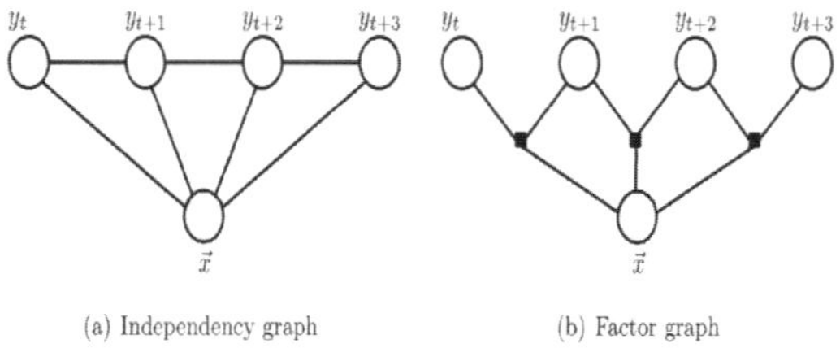

Figure 5: A linear chain Conditional Random Field representation (Klinger and Tomanek, 2007).

In this representation, $\vec{x}$ is a vector of observations, also known as features in machine learning, and the y_t's are states or labels. In this linear chain model, a given state is dependent on its previous, current, and next states. It is also influenced by the observations for that state. This argument can be formulated as Equation 1. Accordingly, state prediction will be an optimization of Equation 1. $\psi_c(\vec{x}, \vec{y})$ are the factor matrices of the maximal cliques read from the factor graph in Figure 5 (Klinger and Tomanek, 2007).

$$P(\vec{y}|\vec{x}) = \frac{1}{Z(\vec{x})} \prod_{c \epsilon C} \psi_c(\vec{x}, \vec{y}) \quad (1)$$

We used the *IOB* format, which is the most common method of representation for sequence tagging. In this format, *I* stands for the inside, *O* is the outside, and *B* is the beginning of a temporal expression. Table 2 shows an example of *IOB* labeling for the phrase ... *late S to G2 phase in HR*.

token	tag
...	...
late	B_TIMEXCCP
S	I_TIMEXCCP
to	I_TIMEXCCP
G2	I_TIMEXCCP
phase	O
in	O
HR	O
.	O

Table 2: IOB format representation of a segment of a sentence.

The rule-based system is keyword-based. The rules match simple cell cycle phase concepts. For example, the phrase *early S phase* is classified as synthesis, since there is *S* in it. The expression *G0/G1 phase* is classified as a *G0/G1* checkpoint.

5 Experimental setup

We split our dataset of more than 70K tokens into 80% training and 20% test sets. We used 5-fold cross validation to balance the distribution of the dataset. The number of positive instances for the 5 runs is shown in Figure 6. The expressions S and synthesis are displayed separately, despite their identical meaning, to allow for more granular evaluation of performance. The same rationale applies to displaying M and mitosis separately.

The ratio of the individual concepts that we have in the 5 runs is balanced, as shown in Figure 6. However, the training dataset is skewed, since there are almost 98% negative labels, with the remaining small portion as positive labels. Among the approximately 10K test tokens, 180 of them are labeled as positive TIMEXCCP, but the others are negative, i.e. they have the label *O*. A positive TIMEXCCP in this case could be B_TIMEXCCP or I_TIMEXCCP—beginning or inside of a temporal expression.,

6 Results

Since the task consisted of two separate steps—temporal expression recognition, and mapping or normalization—in this section, we report our findings independently. Our evaluation metrics are in terms of precision *P*, recall *R*, and *F-measure*. The system achieved precision P = 0.83, recall R = 0.72 and F = 0.77 for recognizing TIMEXCCP in biomedical literature.

The temporal expression mapper, which is a rule-based system, achieved a macro-averaged P= 0.90, R = 0.70, and F = 0.79 and a micro-averaged P = 0.86, R = 0.71, and F = 0.78. The system performance for the individual concepts is shown in Figure 7.

7 Discussion

Some of the false positive predictions were due to human annotation errors.

There were some conditions where the annotators disagreed. For example, ... *early G1 to G2 phase*. This examples addresses two questions that should be explicitly mentioned in the annotation guidelines:

- Does the modifier "early" modify only G1, or both G1 and G2?
- Should there be an attribute for the range of time from G1 to G2 in the annotation guidelines?

Our system achieved good performance on both time expression recognition and mapping of highly ambiguous concepts. In spite of the challenges presented by ambiguity, we obtained 0.85, 0.81, and 0.80 F-measures for recognizing and mapping the concepts *synthesis*, *M*, and *S*, respectively. The most informative features that contribute to this score are the discriminating words before and after a target token. These words are: *phase, arrest*, and *entry*. They are often present before or after CCP concepts. Also, presence of modifiers is a good indication of CCP concepts. For example, in the phase *early S phase*, the modifier *early* is one of the most informative features. However, recognition of complex phrases as in *late S to G2 phase* remained a challenge.

The challenges of complex temporal expressions can be seen from a different perspective.

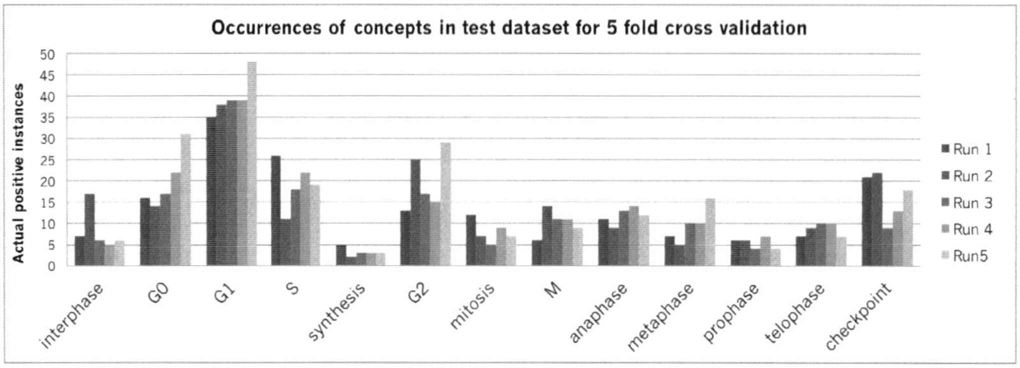

Figure 6: Distribution of concepts in 5 runs.

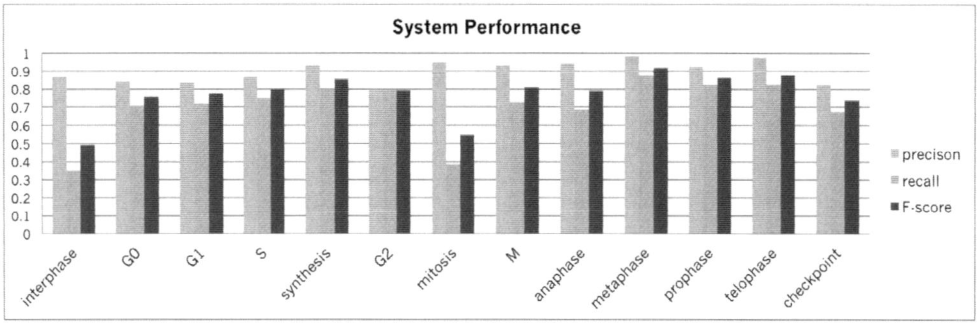

Figure 7: Rule-based classification performance. Average score for 5 runs.

Mostly the system recognizes the individual concepts within a complex phrase, but not the modifiers nor the words like prepositions within the complex phrase. In the example given previously, the system recognizes *S* and *G2* but not the modifier *late*, nor the preposition *to*. These challenges could be tackled by having features that address the modifiers as well as words within two concepts.

We used a naive tokenizer that splits the text into words based on white space. In the future, we would like to test the system with other more sophisticated tokenizers. We kept punctuation marks in temporal expressions, for example, the forward slash in *G0/G1 phase*. Presence of punctuation marks, such as hyphen (-), forward slash (/), comma (,) and single quote ('), within a token is one of our features in training the machine learning algorithm to recognize temporal expressions.

8 Conclusions & Future work

Cell cycle phase concepts are time expressions, and can be annotated in a fashion similar to TimeML. In this work, we annotated a corpus with cell cycle phase information. This corpus can be used to train machine learning algorithms to predict cell cycle phase concepts. The concepts were annotated using the TIMEXCCP tag, an extension of TIMEX3, which has the following attributes: value, modifier, set, and comments. The details are in Figure 3.

We have developed a temporal expression recognizer and classifier based on a hybrid of machine learning and rule-based techniques. We propose a two-tiered architecture to solve temporal expression recognition and mapping for CCP concepts. The first tier recognizes temporal expressions using CRFs. In the second tier, a rule-based system classifies the concepts.

Some of the main future directions for this works are testing the system with the addition of more annotated data. We will focus on how we can capture complex time expressions. This might take us to redefining the annotation guidelines that we have right now.

Acknowledgments

The authors thank Richard Osborne and Scott Cramer for helpful discussion of the significance of this work from a cancer research perspective.

References

Eneko Agirre, Lluís Màrquez, and Richard Wicentowski. 2009. Computational semantic analysis of language: Semeval-2007 and beyond. *Language Resources and Evaluation*, 43(2):97–104.

Daniel Albright, Arrick Lanfranchi, Anwen Fredriksen, William F Styler, Colin Warner, Jena D Hwang, Jinho D Choi, Dmitriy Dligach, Rodney D Nielsen, James Martin, et al. 2013. Towards comprehensive syntactic and semantic annotations of the clinical narrative. *Journal of the American Medical Informatics Association*, 20(5):922–930.

Roberta Alfieri, Ivan Merelli, Ettore Mosca, and Luciano Milanesi. 2007. The Cell Cycle DB: a systems biology approach to cell cycle analysis. *Nucleic Acids Research*.

Juan Carlos Augusto. 2005. Temporal reasoning for decision support in medicine. *Artificial Intelligence in Medicine*, 33(1):1–24.

Matteo Brucato, Leon Derczynski, Hector Llorens, Kalina Bontcheva, and Christian S. Jensen. 2013. Recognising and interpreting named temporal expressions. In Galia Angelova, Kalina Bontcheva, and Ruslan Mitkov, editors, *RANLP*, pages 113–121. RANLP 2011 Organising Committee/ACL.

C. Combi and Y. Shahar. 1997. Temporal reasoning and temporal data maintenance in medicine: Issues and challenges. *Comput Biol Med*, 27 (5).

Carlo Combi, Elpida Keravnou-Papailiou, and Yuval Shahar. 2010. *Temporal Information Systems in Medicine*. Springer Publishing Company, Incorporated, 1st edition.

Collins Editorial Staff. undated. *Collins Concise English Dictionary*.

Nicholas Paul Gauthier, Lars Juhl Jensen, Rasmus Wernersson, Sören Brunak, and Thomas S. Jensen. 2009. Cyclebase.org: version 2.0, an updated comprehensive, multi-species repository of cell cycle experiments and derived analysis results. *Nucleic Acids Research*, 9.

Erik Hatcher, Otis Gospodnetic, and Mike McCandless. 2nd revised edition. edition.

George Hripcsak and Adam S Rothschild. 2005. Agreement, the f-measure, and reliability in information retrieval. *Journal of the American Medical Informatics Association*, 12(3):296–298.

Lawrence Hunter and K. Bretonnel Cohen. 2006. Biomedical Language Processing: Perspective What's Beyond PubMed? *Molecular Cell*, 21:589–594.

Michael Kahn. December 1991. Modeling time in medical decision-support programs. *Med Decision Making*, 11(4):249–264.

Michael B. Kastan and Jiri Bartek. 2004. Cell-cycle checkpoints and cancer. *Nature*, 432:316–323.

Fergal C. Kelleher, Aparna Rao, and Anne Maguire. 2014. Circadian molecular clocks and cancer. *Cancer Letters*, 342:9–18.

Roman Klinger and Katrin Tomanek. 2007. Classical Probabilistic Models and Conditional Random Fields. Technical Report TR07-2-013, Department of Computer Science, Dortmund University of Technology, December.

R. Leaman and Gonzalez G. 2008. BANNER: An executable survey of advances in biomedical named entity recognition. *Pacific Symposium on Biocomputing*, 13:652–663.

Zsuzsanna Lengyel, Zita Battyáni, György Szekeres, Valér Csernus, and András D. Nagy. 2009. Circadian clocks and tumor biology: what is to learn from human skin biopsies? *Nature Reviews Cancer*, 9:153–166.

Marcos Malumbres and Mariano Barbacid. 2009. Cell Cycle, CDKs and cancer: a changing paradigm. *Nature Reviews Cancer*, 9:153–166.

Inderjeet Mani, Ben Wellner, Marc Verhagen, and James Pustejovsky. 2007. Three approaches to learning TLINKs in TimeML. Technical report, Computer Science Department, Brandeis University. Waltham, USA.

Christopher D. Manning, Prabhakar Raghavan, and Hinrich Schütze. 2008. *Introduction to Information Retrieval*. Cambridge University Press, Cambridge, UK.

Andrew Kachites McCallum. 2002. MALLET: A machine learning for language toolkit.

David Nadeau and Satoshi Sekine. 2007. A survey of named entity recognition and classification. *Linguisticae Investigationes*, 30(1):3–26, January. Publisher: John Benjamins Publishing Company.

Philip V. Ogren. 2006. Knowtator: a protégé plug-in for annotated corpus construction. In *Proceedings of the 2006 Conference of the North American Chapter of the Association for Computational Linguistics on Human Language Technology*, pages 273–275, Morristown, NJ, USA. Association for Computational Linguistics.

Elisabet Ortiz-Tudela, Ida Iurisci, Jacques Beau, Abdoulaye Karaboue, Thierry Moreau, Maria Angeles Rol, Juan Antonio Madrid, Francis Lévi, and Pasquale F. Innominato. 2013. The circadian rest-activity rhythm, a potential safety pharmacology endpoint of cancer chemotherapy. *International Journal of Cancer*.

Martha Palmer and James Pustejovsky. 2012. 2012 i2b2 temporal relations challenge annotation guidelines.

James Pustejovsky and Amber Stubbs. 2012. *Natural Language Annotation for Machine Learning*. O'REILLY.

James Pustejovsky, Josè Castano, Robert Ingria, Roser Saur, Robert Gaizauskas, Andrea Setzer, and Graham Katz. 2003. TimeML: Robust specification of event and temporal expressions in text. In *Fifth International Workshop on Computational Semantics (IWCS-5)*.

Saurabh Sahar and Paolo Sassone-Corsi. 2009. Metabolism and cancer: the circadian clock connection. *Nature*, 9:886–896.

Yuval Shahar and Carlo Combi. 1999. Editors' foreword: Intelligent temporal information systems in medicine. *J. Intell. Inf. Syst.*, 13(1-2):5–8.

Paul T. Spellman, Gavin Sherlock, Michael Q. Zhang, Vishwanath R. Iyer, Kirk Anders, Michael B. Eisen, Patrick O. Brown, David Botstein, and Bruce Futcher. 1998. Comprehensive identification of cell cycle-regulated genes of the yeast Saccharomyces cerevisiae by microarray hybridization. *Molecular Biology of the Cell*, 9.

William F Styler IV, Steven Bethard, Sean Finan, Martha Palmer, Sameer Pradhan, Piet C de Groen, Brad Erickson, Timothy Miller, Chen Lin, and Guergana Savova. Undated. Temporal annotation in the clinical domain.

Naushad UzZaman, Hector Llorens, Leon Derczynski, James Allen, Marc Verhagen, and James Pustejovsky. 2013. Semeval-2013 task 1: Tempeval-3: Evaluating time expressions, events, and temporal relations. In *Second Joint Conference on Lexical and Computational Semantics (*SEM), Volume 2: Proceedings of the Seventh International Workshop on Semantic Evaluation (SemEval 2013)*, pages 1–9, Atlanta, Georgia, USA, June. Association for Computational Linguistics.

Katrien Vermeulen, Dirk R. Van Bockstaele, and Zwi N. Berneman. 2003. The Cell Cycle: a review of regulation, deregulation and therapeutic targets in cancer. *Cell Proliferation*, 36:131–149.

Michael L. Whitield, Gavin Sherlock, Alok J. Saldanha, John I. Murray, Catherine A. Ball, Karen E. Alexander, John C. Matese, Charles M. Perou, Myra M. Hurt, Patrick O. Brown, and David Botstein. 2002. Identication of genes periodically expressed in the human cell cycle and their expression in tumors. *Molecular Biology of the Cell*, 13.

Li Zhou and George Hripcsak. 2007. Temporal reasoning with medical data - a review with emphasis on medical natural language processing. *Journal of Biomedical Informatics*, 40(2):183–202.

Classifying Negative Findings in Biomedical Publications

Bei Yu
School of Information Studies
Syracuse University
byu@syr.edu

Daniele Fanelli
School of Library and Information Science
University of Montreal
email@danielefanelli.com

Abstract

Publication bias refers to the phenomenon that statistically significant, "positive" results are more likely to be published than non-significant, "negative" results. Currently, researchers have to manually identify negative results in a large number of publications in order to examine publication biases. This paper proposes an NLP approach for automatically classifying negated sentences in biomedical abstracts as either reporting negative findings or not. Using multinomial naïve Bayes algorithm and bag-of-words features enriched by parts-of-speeches and constituents, we built a classifier that reached 84% accuracy based on 5-fold cross validation on a balanced data set.

1 Introduction

Publication bias refers to the phenomenon that statistically significant, "positive" results are more likely to be published than non-significant, "negative" results (Estabrook et al., 1991). Due to the "file-drawer" effect (Rosenthal, 1979), negative results are more likely to be "filed away" privately than to be published publicly.

Publication bias poses challenge for an accurate review of current research progress. It threatens the quality of meta-analyses and systematic reviews that rely on published research results (e.g., the Cochrane Review). Publication bias may be further spread through citation network, and amplified by citation bias, a phenomenon that positive results are more likely to be cited than negative results (Greenberg, 2009).

To address the publication bias problem, some new journals were launched and dedicated to publishing negative results, such as the Journal of Negative Results in Biomedicine, Journal of Pharmaceutical Negative Results, Journal of Negative Results in Ecology and Evolutionary Biology, and All Results Journals: Chem. Some quantitative methods like the funnel plot (Egger et al., 1997) were used to measure publication bias in publications retrieved for a certain topic.

A key step in such manual analysis is to examine the article abstracts or full-texts to see whether the findings are negative or not. For example, Hebert et al. (2002) examined the full text of 1,038 biomedical articles whose primary outcomes were hypothesis testing results, and found 234 (23%) negative articles. Apparently, such manual analysis approach is time consuming. An accurate, automated classifier would be ideal to actively track positive and negative publications.

This paper proposes an NLP approach for automatically identifying negative results in biomedical abstracts. Because one publication may have multiple findings, we currently focus on classifying negative findings at sentence level: for a sentence that contains the negation cues "no" and/or "not", we predict whether the sentence reported negative finding or not. We constructed a training data set using manual annotation and convenience samples. Two widely-used text classification algorithms, Multinomial naive Bayes (MNB) and Support Vector Machines (SVM), were compared in this study. A few text representation approaches were also compared by their effectiveness in building the classifier. The approaches include (1) bag-of-words (BOW), (2) BOW with PoS tagging and shallow parsing, and (3) local contexts of the negation cues "no" and 'not", including the words, PoS tags, and constituents. The best classifier was built using MNB and bag-of-words features enriched with PoS tags and constituent markers. The best performance is 84% accuracy based on 5-fold cross validation on a balanced data set.

Proceedings of the 2014 Workshop on Biomedical Natural Language Processing (BioNLP 2014), pages 19–23,
Baltimore, Maryland USA, June 26-27 2014.

2 Related work

The problem of identifying negative results is related to several other BioNLP problems, especially on negation and scientific claim identification.

The first relevant task is to identify negation signals and their scopes (e.g., Morante and Daelemans, 2008;2009; Farkas et al., 2010; Agarwal et al., 2011). Manually-annotated corpora like BioScope (Szarvas et al., 2008) were created to annotate negations and their scopes in biomedical abstracts in support of automated identification. This task targets a wide range of negation types, such as the presence or absence of clinical observations in narrative clinical reports (Chapman et al., 2001). In comparison, our task focuses on identifying negative findings only. Although not all negations report negative results, negation signals are important rhetorical device for authors to make negative claims. Therefore, in this study we also examine precision and recall of using negation signals as predictors of negative findings.

The second relevant task is to identify the strength and types of scientific claims. Light et al. (2004) developed a classifier to predict the level of speculations in sentences in biomedical abstracts. Blake (2010) proposed a "claim framework" that differentiates explicit claims, observations, correlations, comparisons, and implicit claims, based on the certainty of the causal relationship that was presented. Blake also found that abstracts contained only 7.84% of all scientific claims, indicating the need for full-text analysis. Currently, our preliminary study examines abstracts only, assuming the most important findings are reported there. We also focus on coarse-grained classification of positive vs. negative findings at this stage, and leave for future work the task of differentiating negative claims in finer-granularity.

3 The NLP approach

3.1 The definition of negative results

When deciding what kinds of results count as "negative", some prior studies used "non-significant" results as an equivalent for "negative results" (e.g. Hebert et al., 2002; Fanelli, 2012). However, in practice, the definition of "negative results" is actually broader. For example, the Journal of Negative Results in Biomedicine (JNRBM), launched in 2002, was devoted to publishing "unexpected, controversial, provocative and/or negative results," according to the journal's website. This broader definition has its pros and cons. The added ambiguity poses challenge for manual and automated identification. At the same time, the broader definition allows the inclusion of descriptive studies, such as the first JNRBM article (Hebert et al., 2002).

Interestingly, Hebert et al. (2002) defined "negative results" as "non-significant outcomes" and drew a negative conclusion that "prominent medical journals often provide insufficient information to assess the validity of studies with negative results", based on descriptive statistics, not hypothesis testing. This finding would not be counted as "negative" unless the broader definition is adopted.

In our study, we utilized the JNRBM articles as a convenience sample of negative results, and thus inherit its broader definition.

3.2 The effectiveness of negation cues as predictors

The Bioscope corpus marked a number of negation cues in the abstracts of research articles, such as "not", "no", "without", etc. It is so far the most comprehensive negation cue collection we can find for biomedical publications. However, some challenges arise when applying these negation cues to the task of identifying negative results.

First, instead of focusing on negative results, the Bioscope corpus was annotated with cues expressing general negation and speculations. Consequently, some negation cues such as "unlikely" was annotated as a speculation cue, not a negation cue, although "unlikely" was used to report negative results like

> "*These data indicate that changes in Wnt expression per se are* ***unlikely*** *to be the cause of the observed dysregulation of β-catenin expression in DD*" (PMC1564412).

Therefore, the Bioscope negation cues may not have captured all negation cues for reporting negative findings. To test this hypothesis, we used the JNRBM abstracts (N=90) as a convenience sample of negative results, and found that 81 abstracts (88.9%) contain at least one Bioscope negation cue. Note that because the JNRBM abstracts consist of multiple subsections "background", "objective", "method", "result", and "conclusion", we used the "result" and "conclusions" subsections only to narrow down the search range.

Among the 9 missed abstracts, 5 used cues not captured in Bioscope negation cues: "insufficient", "unlikely", "setbacks", "worsening", and "underestimates". However, the authors' writing style might be affected by the fact that JNRBM is dedicated to negative results. One hypothesis is that the authors would feel less pressure to use negative tones, and thus used more variety of negation words. Hence we leave it as an open question whether the new-found negation cues and their synonyms are generalizable to other biomedical journal articles.

The rest 4 abstracts (PMC 1863432, 1865554, 1839113, and 2746800) did not report explicit negation results, indicating that sometimes abstracts alone are not enough to decide whether negative results were reported, although the percentage is relatively low (4.4%). Hence, we decided that missing target is not a major concern for our task, and thus would classify a research finding as positive if no negation cues were found.

Second, some positive research results may be mistaken as negative just because they used negation cues. For example, "without" is marked as a negation cue in Bioscope, but it can be used in many contexts that do not indicate negative results, such as

> *"The effects are consistent with or without the presence of hypertension and other comorbidities and across a range of drug classes."* (PMC2659734)

To measure the percentage of false alarm, we applied the aforementioned trivial classifier to a corpus of 600 abstracts in 4 biomedical disciplines, which were manually annotated by Fanelli (2012). This corpus will be referred to as "Corpus-600" hereafter. Each abstract is marked as "positive", "negative", "partially positive", or "n/a", based on hypothesis testing results. The latter two types were excluded in our study. The trivial classifier predicted an abstract as "positive" if no negation cues were found. Table 1 reported the prediction results, including the precision and recall in identifying negative results. This result corroborates with our previous finding that the inclusiveness of negation cues is not the major problem since high recalls have been observed in both experiments. However, the low precision is the major problem in that the false negative predictions are far more than the true negative predictions. Hence, weeding out the negations that did not report negative results became the main purpose of this preliminary study.

Discipline	#abstracts	Precision	Recall
Psychiatry	140	.11	.92
Clinical Medicine	127	.16	.94
Neuroscience	144	.20	.95
Immunology	140	.18	.95
Total	551	.16	.94

Table 1: results of cue-based trivial classifier

3.3 Classification task definition

This preliminary study focuses on separating negations that reported negative results and those not. We limit our study to abstracts at this time. Because a paper may report multiple findings, we performed the prediction at sentence level, and leave for future work the task of aggregating sentence-level predictions to abstract-level or article-level. By this definition, we will classify each sentence as reporting negative finding or not. A sentence that includes mixed findings will be categorized as reporting negative finding.

"Not" and "no" are the most frequent negation cues in the Bioscope corpus, accounting for more than 85% of all occurrences of negation cues. In this study we also examined whether local context, such as the words, parts-of-speeches, and constituents surrounding the negation cues, would be useful for predicting negative findings. Considering that different negation cues may be used in different contexts to report negative findings, we built a classifier based on the local contexts of "no" and "not". Contexts for other negation cues will be studied in the future.

Therefore, our goal is to extract sentences containing "no" or "not" from abstracts, and predict whether they report negative findings or not.

3.4 Training data

We obtained a set of "positive examples", which are negative-finding sentences, and a set of "negative examples" that did not report negative findings. The examples were obtained in the following way.

Positive examples. These are sentences that used "no" or "not" to report negative findings. We extracted all sentences that contain "no" or "not" in JNRBM abstracts, and manually marked each sentence as reporting negative findings or not. Finally we obtained 158 sentences reporting negative findings.

To increase the number of negative-finding examples and add variety to writing styles, we repeat the above annotations to all Lancet abstracts ("result" and "finding" subsections only) in the PubMed Central open access subset, and obtained 55 more such sentences. Now we have obtained 213 negative-finding examples in total.

Negative examples. To reduce the workload for manual labeling, we utilized the heuristic rule that a "no" or "not" does not report negative result if it occurs in a positive abstract, therefore we extracted such sentences from positive abstracts in "Corpus-600". These are the negative examples we will use. To balance the number of positive and negative examples, we used a total of 231 negative examples in two domains (132 in clinical medicine and 99 in neuroscience) instead of all four domains, because there are not enough positive examples.

Now the training data is ready for use.

3.5 Feature extraction

We compared three text representation methods by their effectiveness in building the classifier. The approaches are (1) BOW: simple bag-of-words, (2) E-BOW: bag-of-words enriched with PoS tagging and shallow parsing, and (3) LCE-BOW: local contexts of the negation cues "no" and 'not", including the words, PoS tags, and constituents. For (2) and (3), we ran the OpenNLP chunker through all sentences in the training data. For (3), we extracted the following features for each sentence:

- The type of chunk (constituent) where "no/not" is in (e.g. verb phrase "VP");
- The types of two chunks before and after the chunk where "not" is in;
- All words or punctuations in these chunks;
- The parts-of-speech of all these words.

See Table 2 below for an example of negative finding: row 1 is the original sentence; row 2 is the chunked sentence, and row 3 is the extracted local context of the negation cue "not". These three representations were then converted to feature vectors using the "bag-of-words" representation. To reduce vocabulary size, we removed words that occurred only once.

(1)	Vascular mortality did not differ significantly (0.19% vs 0.19% per year, p=0.7).
(2)	"[NP Vascular/JJ mortality/NN] [VP did/VBD not/RB differ/VB] [ADVP significantly/RB] [PP (/-LRB-] [NP 019/CD %/NN] [PP vs/IN] [NP 019/CD %/NN] [PP per/IN] [NP year/NN] ,/, [NP p=07/NNS] [VP)/-RRB-] ./."
(3)	"na na VP ADVP PP did not differ significantly VBD RB VB RB"

Table 2: text representations

3.6 Classification result

We applied two supervised learning algorithms, multinomial naïve Bayes (MNB), and Support Vector Machines (Liblinear) to the unigram feature vectors. We used the Sci-kit Learn toolkit to carry out the experiment, and compared the algorithms' performance using 5-fold cross validation. All algorithms were set to the default parameter setting.

Representation		MNB	SVM
Presence vs. absence	BOW	.82	.79
	E-BOW	.82	.79
	LCE-BOW	.72	.72
tf	BOW	.82	.79
	E-BOW	.84	.79
	LCE-BOW	.72	.72
Tfidf	BOW	.82	.75
	E-BOW	**.84**	.73
	LCE-BOW	.72	.75

Table 3: classification accuracy

Table 3 reports the classification accuracy. Because the data set contains 213 positive and 231 negative examples, the majority vote baseline is .52. Both algorithms combined with any text representation methods outperformed the majority baseline significantly. Among them the best classifier is a MNB classifier based on enriched bag-of-words representation and tfidf weighting. Although LCE-BOW reached as high as .75 accuracy using SVM and tfidf weighting, it did not perform as well as the other text representation methods, indicating that the local context with +/- 2 window did not capture all relevant indicators for negative findings.

Tuning the regularization parameter C in SVM did not improve the accuracy. Adding bigrams to the feature set resulted in slightly lower accuracy.

4 Conclusion

In this study we aimed for building a classifier to predict whether a sentence containing the words "no" or "not" reported negative findings. Built with MNB algorithms and enriched bag-of-words features with tfidf weighting, the best classifier reached .84 accuracy on a balanced data set.

This preliminary study shows promising results for automatically identifying negative findings for the purpose of tracking publication bias. To reach this goal, we will have to aggregate the sentence-level predictions on individual findings to abstract- or article-level negative results. The aggregation strategy is dependent on the decision of which finding is the primary outcome when multiple findings are present. We leave this as our future work.

Reference

S. Agarwal, H. Yu, and I. Kohane, I. 2011. BioNOT: A searchable database of biomedical negated sentences. BMC bioinformatics, 12: 420.

C. Blake. 2010. Beyond genes, proteins, and abstracts: Identifying scientific claims from full-text biomedical articles. *Journal of biomedical informatics*, 43(2): 173-189.

W. W. Chapman, W. Bridewell, P. Hanbury, G. F. Cooper, and B. G. Buchanan. 2001. Evaluation of negation phrases in narrative clinical reports. *Proceedings of the AMIA Symposium*, 105.

P. J. Easterbrook, R. Gopalan, J. A. Berlin, and D. R. Matthews. 1991. Publication bias in clinical research. *Lancet*, 337(8746): 867-872.

M. Egger, G. D. Smith, M. Schneider, and C. Minder. 1997. Bias in meta-analysis detected by a simple, graphical test. *BMJ* 315(7109): 629-634.

D. Fanelli. 2012. Negative results are disappearing from most disciplines and countries. *Scientometrics* 90(3): 891-904.

R. Farkas, V. Vincze, G. Móra, J. Csirik, and G. Szarvas. 2010. The CoNLL-2010 shared task: learning to detect hedges and their scope in natural language text. *Proceedings of the Fourteenth Conference on Computational Natural Language Learning---Shared Task*, 1-12.

S. A. Greenberg. 2009. How citation distortions create unfounded authority: analysis of a citation network. *BMJ* 339, b2680.

R. S. Hebert, S. M. Wright, R. S. Dittus, and T. A. Elasy. 2002. Prominent medical journals often provide insufficient information to assess the validity of studies with negative results. *Journal of Negative Results in Biomedicine* 1(1):1.

M. Light, X-Y Qiu, and P. Srinivasan. 2004. The language of bioscience: Facts, speculations, and statements in between. *In Proceedings of BioLink 2004 workshop on linking biological literature, ontologies and databases: tools for users*, pp. 17-24.

R. Morante, A. Liekens, and W. Daelemans. 2008. Learning the scope of negation in biomedical texts. *In Proceedings of the Conference on Empirical Methods in Natural Language Processing*, pp. 715-724.

R. Morante, and W. Daelemans. 2009. A metalearning approach to processing the scope of negation. *Proceedings of the Thirteenth Conference on Computational Natural Language Learning*, 21-29.

R. Rosenthal. 1979. The file drawer problem and tolerance for null results. *Psychological Bulletin*, 86(3): 638.

G. Szarvas, V. Vincze, R. Farkas, and J. Csirik. 2008. The BioScope corpus: annotation for negation, uncertainty and their scope in biomedical texts. *In Proceedings of the Workshop on Current Trends in Biomedical Natural Language Processing,* pp. 38-45.

Automated Disease Normalization with Low Rank Approximations

Robert Leaman **Zhiyong Lu**
National Center for Biotechnology Information
National Library of Medicine
{robert.leaman, zhiyong.lu}@nih.gov

Abstract

While machine learning methods for named entity recognition (mention-level detection) have become common, machine learning methods have rarely been applied to normalization (concept-level identification). Recent research introduced a machine learning method for normalization based on pairwise learning to rank. This method, DNorm, uses a linear model to score the similarity between mentions and concept names, and has several desirable properties, including learning term variation directly from training data. In this manuscript we employ a dimensionality reduction technique based on low-rank matrix approximation, similar to latent semantic indexing. We compare the performance of the low rank method to previous work, using disease name normalization in the NCBI Disease Corpus as the test case, and demonstrate increased performance as the matrix rank increases. We further demonstrate a significant reduction in the number of parameters to be learned and discuss the implications of this result in the context of algorithm scalability.

1 Introduction

The data necessary to answer a wide variety of biomedical research questions is locked away in narrative text. Automating the location (named entity recognition) and identification (normalization) of key biomedical entities (Doğan et al., 2009; Névéol et al., 2011) such as diseases, proteins and chemicals in narrative text may reduce curation costs, enable significantly increased scale and ultimately accelerate biomedical discovery (Wei et al., 2012a).

Named entity recognition (NER) techniques have typically focused on machine learning methods such as conditional random fields (CRFs), which have provided high performance when coupled with a rich feature approach. The utility of NER for biomedical end users is limited, however, since many applications require each mention to be normalized, that is, identified within a specified controlled vocabulary.

The normalization task has been highlighted in the BioCreative challenges (Hirschman et al., 2005; Lu et al., 2011; Morgan et al., 2008), where a variety of methods have been explored for normalizing gene names, including string matching, pattern matching, and heuristic rules. Similar methods have been applied to disease names (Doğan & Lu, 2012b; Kang et al., 2012; Névéol et al., 2009) and species names (Gerner et al., 2010; Wei et al., 2012b), and the MetaMap program is used to locate and identify concepts from the UMLS MetaThesaurus (Aronson, 2001; Bodenreider, 2004).

Machine learning methods for NER have provided high performance, enhanced system adaptability to new entity types, and abstracted many details of specific rule patterns. While machine learning methods for normalization have been explored (Tsuruoka et al., 2007; Wermter et al., 2009), these are far less common. This is partially due to the lack of appropriate training data, and also partially due to the need for a generalizable supporting framework.

Normalization is frequently decomposed into the sub-tasks of candidate generation and disambiguation (Lu et al., 2011; Morgan et al., 2008). During candidate generation, the set of concept names is constrained to a set of possible matches using the text of the mention. The primary difficulty addressed in candidate generation is term variation: the need to identify terms which are semantically similar but textually distinct (e.g. "nephropathy" and "kidney disease"). The disambiguation step then differentiates between the different candidates to remove false positives, typically using the context of the mention and the article metadata.

Proceedings of the 2014 Workshop on Biomedical Natural Language Processing (BioNLP 2014), pages 24–28,
Baltimore, Maryland USA, June 26-27 2014.

Recently, Leaman et al. (2013a) developed an algorithm (DNorm) that directly addresses the term variation problem with machine learning, and used diseases – an important biomedical entity – as the first case study. The algorithm learns a similarity function between mentions and concept names directly from training data using a method based on pairwise learning to rank. The method was shown to provide high performance on the NCBI Disease Corpus (Doğan et al., 2014; Doğan & Lu, 2012a), and was also applied to clinical notes in the ShARe / CLEF eHealth task (Suominen et al., 2013), where it achieved the highest normalization performance out of 17 international teams (Leaman et al., 2013b). The normalization step does not consider context, and therefore must be combined with a disambiguation method for tasks where disambiguation is important. However, this method provides high performance when paired with a conditional random field system for NER, making the combination a step towards fully adaptable mention recognition and normalization systems.

This manuscript adapts DNorm to use a dimensionality reduction technique based on low rank matrix approximation. This may provide several benefits. First, it may increase the scalability of the method, since the number of parameters used by the original technique is proportional to the square of the number of unique tokens. Second, reducing the number of parameters may, in turn, improve the stability of the method and improve its generalization due to the induction of a latent "concept space," similar to latent semantic indexing (Bai et al., 2010). Finally, while the rich feature approach typically used with conditional random fields allows it to partially compensate for out-of-vocabulary effects, DNorm ignores unknown tokens. This reduces the ability of the model to generalize, due to the zipfian distribution of text (Manning & Schütze, 1999), and is especially problematic in text which contains many misspellings, such as consumer text. Using a richer feature space with DNorm would not be feasible, however, unless the parameter scalability problem is resolved.

In this article we expand the DNorm method in a pilot study on feasibility of using low rank approximation methods for disease name normalization. To make this work comparable to the previous work on DNorm, we again employed the NCBI Disease Corpus (Doğan et al., 2014). This corpus contains nearly 800 abstracts, split into training, development, and test sets, as described in Table 1. Each disease mention is annotated for span and concept, using the MEDIC vocabulary (Davis et al., 2012), which combines MeSH® (Coletti & Bleich, 2001) and OMIM® (Amberger et al., 2011). The average number of concepts for each name in the vocabulary is 5.72. Disease names exhibit relatively low ambiguity, with an average number of concepts per name of 1.01.

Subset	Abstracts	Mentions	Concepts
Training	593	5145	670
Development	100	787	176
Test	100	960	203

Table 1. Descriptive statistics for the NCBI Disease Corpus.

2 Methods

DNorm uses the BANNER NER system (Leaman & Gonzalez, 2008) to locate disease mentions, and then employs a ranking method to normalize each mention found to the disease concepts in the lexicon (Leaman et al., 2013a). Briefly, we define $\mathcal{T}$ to be the set of tokens from both the disease mentions in the training data and the concept names in the lexicon. We stem each token in both disease mentions and concept names (Porter, 1980), and then convert each to TF-IDF vectors of dimensionality $|\mathcal{T}|$, where the document frequency for each token is taken to be the number of names in the lexicon containing it (Manning et al., 2008). All vectors are normalized to unit length. We define a similarity score between mention vector m and name vector n, $score(m, n)$, and each mention is normalized by iterating through all concept names and returning the disease concept corresponding to the one with the highest score.

In previous work, $score(m, n) = m^{\mathsf{T}} W n$, where W is a weight matrix and each entry w_{ij} represents the correlation between token t_i appearing in a mention and token t_j appearing in a concept name from the lexicon. In this work, however, we set W to be a low-rank approximation of the form $W = U^{\mathsf{T}} V + I$, where U and V are both $r \times |\mathcal{T}|$ matrices, r being the rank (number of linearly independent rows), and $r \ll |\mathcal{T}|$ (Bai et al., 2010).

For efficiency, the low-rank scoring function can be rewritten and evaluated as $score(m, n) = (Um)^{\mathsf{T}}(Vn) + m^{\mathsf{T}} n$, allowing the respective Um and Vn vectors to be calculated once and then reused. This view provides an intuitive explanation of the purpose of the U and V matrices: to convert the sparse, high-dimensional mention and concept name vectors (m and n) into dense, low dimensional vectors (as Um and Vn). Under this interpretation, we found that performance improved if each Um and Vn vector was renormalized to unit length.

This model retains many useful properties of the original model, such as the ability to represent both positive and negative correlations between tokens, to represent both synonymy and polysemy, and to allow the token distributions between the mentions and the names to be different. The new model also adds one important additional property: the number of parameters is linear in the number of unique tokens, potentially enabling greater scalability.

2.1 Model Training

Given any pair of disease names where one (n^+) is for c^+ , the correct disease concept for tion m, and the other, n^-, is for c^-, an incorrect concept , we would like to update the weight matrix W so that $m^\mathsf{T}Wn^+ > m^\mathsf{T}Wn^-$. Following Leaman et al. (2013a), we iterate through each $\langle m, c^+, c^- \rangle$ tuple, selecting $n^+$and n^-as the name for c^+ and c^-, respectively, with the highest similarity score to m, using stochastic gradient descent to make updates to W. With a dense weight matrix W , the update rule is: if $m^\mathsf{T}Wn^+ - m^\mathsf{T}Wn^- < 1$, then W is updated as $W \leftarrow W + \eta(m(n^+)^\mathsf{T} - m(n^-)^\mathsf{T})$, where η is the learning rate, a parameter controlling the size of the change to W. Under the low-rank approximation, the update rules are: if $m^\mathsf{T}Wn^+ - m^\mathsf{T}Wn^- < 1$, then U is updated as $U \leftarrow U + \eta V(n^+ - n^-)m^\mathsf{T}$, and V is updated as $V \leftarrow V + \eta Um(n^+ - n^-)^\mathsf{T}$, noting that the updates are applied simultaneously (Bai et al., 2010). Overfitting is avoided using a holdout set, using the average of the ranks of the correct concept as the performance measurement, as in previous work.

We initialize U using values chosen randomly from a normal distribution with mean 0 and standard deviation 1. We found it useful to initialize V as U^T, since this causes the representation for disease mentions and disease names to initially be the same.

We employed an adaptive learning rate using the schedule $\eta_k = \eta_0 \frac{\tau}{\tau+k}$, where k is the iteration, η_0 is the initial learning rate, and τ is the discount (Finkel et al., 2008). We used an initial learning rate of $\eta_0 = 10^{-7}$. This is much lower than reported by Leaman et al. (2013a), since we found that higher values caused the training to diverge. We used a discount parameter of $\tau = 5$, so that the learning rate is equal to one half the initial rate after five iterations.

3 Results

Our results were evaluated at the abstract level, allowing comparison to the previous work on DNorm (Leaman et al., 2013a). This evaluation considers the set of disease concepts found in the abstract, and ignores the exact location(s) where each concept was found. A true positive consists of the system returning a disease concept annotated within the NCBI Disease Corpus, and the number of false negatives and false positives are defined similarly. We calculated the precision, recall and F-measure as follows:

$$p = \frac{tp}{tp + fp} \quad r = \frac{tp}{tp + fn} \quad f = \frac{2pr}{p + r}$$

We list the micro-averaged results in Table 2.

Rank	Precision	Recall	F-measure
50	0.648	0.671	0.659
100	0.673	0.685	0.679
250	0.697	0.697	0.697
500	0.702	0.700	0.701
(Full)	0.828	0.819	0.809

Table 2. Performance measurements for each model on the NCBI Disease Test set. Full corresponds with the full-rank matrix used in previous work.

4 Discussion

There are two primary trends to note. First, the performance of the low rank models is about 10%-15% lower than the full rank model. Second, there is a clear trend towards higher precision and recall as the rank of the matrix increases. This trend is reinforced in Figure 1, which shows the learning curve for all models. These describe the performance on the holdout set after each iteration through the training data, and are measured using the average rank of the correct concept in the holdout set, which is dominated by a small number of difficult cases.

Using the low rank approximation, the number of parameters is equal to $2 \times r \times |\mathcal{T}|$. Since r is fixed and independent of $|\mathcal{T}|$, the number of parameters is now linear in the number of tokens, effectively solving the parameter scalability problem. Table 3 lists the number of parameters for each of the models used in this study.

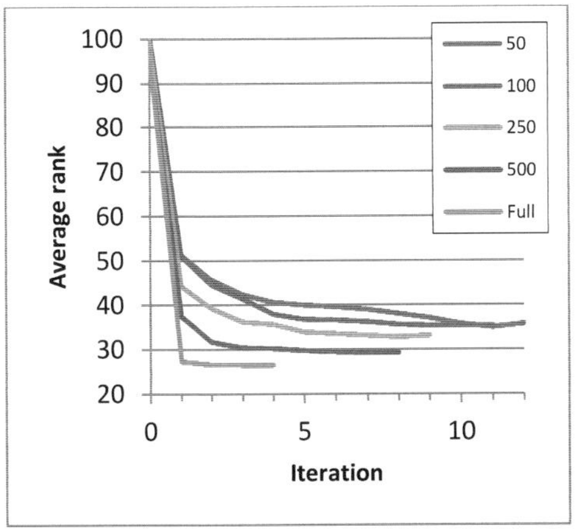

Figure 1. Learning curves showing holdout performance at each iteration through the training data.

Rank	Parameters
50	1.8×10^6
100	3.7×10^6
250	9.1×10^6
500	1.8×10^7
(Full)	3.3×10^8

Table 3. Number of model parameters for each variant, showing the low rank methods using 1 to 2 orders of magnitude fewer parameters.

There are two trade-offs for this improvement in scalability. First, there is a substantial performance reduction, though this might be mitigated somewhat in the future by using a richer feature set – a possibility enabled by the use of the low rank approximation. Second, training and inference times are significantly increased; training the largest low-rank model ($r = 500$) required approximately 9 days, though the full-rank model trains in under an hour.

The view that the *U* and *V* matrices convert the TF-IDF vectors to a lower dimensional space suggests that the function of *U* and *V* is to provide word embeddings or word representations – a vector space where each word vector encodes its relationships with other words. This further suggests that one way to provide higher performance may be to take advantage of unsupervised pre-training (Erhan et al., 2010). Instead of initializing *U* and *V* randomly, they could be initialized using a set of word embeddings trained on a large amount of biomedical text, such as with neural network language models (Collobert & Weston, 2008; Mikolov et al., 2013).

5 Conclusion

We performed a pilot study to determine whether a low rank approximation may increase the scalability of normalization using pairwise learning to rank. We showed that the reduction in the number of parameters is substantial: it is now linear to the number of tokens, rather than proportional to the square of the number of tokens. We further observed that the precision and recall increase as the rank of the matrices is increased.

We believe that further performance increases may be possible through the use of a richer feature set, unsupervised pre-training, or other dimensionality reduction techniques including feature selection or L_1 regularization (Tibshirani, 1996). We also intend to apply the method to additional entity types, using recently released corpora such as CRAFT (Bada et al., 2012).

Acknowledgments

The authors would like to thank the anonymous reviewers for their helpful suggestions. This research was supported by the NIH Intramural Research Program, National Library of Medicine.

References

Amberger, J., Bocchini, C., & Hamosh, A. (2011). A new face and new challenges for Online Mendelian Inheritance in Man (OMIM(R)). *Hum Mutat, 32*(5), 564-567.

Aronson, A. R. (2001). *Effective mapping of biomedical text to the UMLS Metathesaurus: the MetaMap program*. In Proceedings of the AMIA Symposium, 17-21.

Bada, M., Eckert, M., Evans, D., Garcia, K., Shipley, K., Sitnikov, D., et al. (2012). Concept annotation in the CRAFT corpus. *BMC Bioinformatics, 13*, 161.

Bai, B., Weston, J., Grangier, D., Collobert, R., Sadamasa, K., Qi, Y. J., et al. (2010). Learning to rank with (a lot of) word features. *Inform. Retrieval, 13*(3), 291-314.

Bodenreider, O. (2004). The Unified Medical Language System (UMLS): integrating biomedical terminology. *Nucleic Acids Res, 32*, D267-270.

Coletti, M. H., & Bleich, H. L. (2001). Medical subject headings used to search the biomedical literature. *J Am Med Inform Assoc, 8*(4), 317-323.

Collobert, R., & Weston, J. (2008). *A unified architecture for natural language processing: deep neural networks with multitask learning*. In Proceedings of the ICML, 160-167.

Davis, A. P., Wiegers, T. C., Rosenstein, M. C., & Mattingly, C. J. (2012). MEDIC: a practical disease vocabulary used at the Comparative

Toxicogenomics Database. *Database, 2012*, bar065.
Doğan, R. I., Leaman, R., & Lu, Z. (2014). NCBI disease corpus: A resource for disease name recognition and concept normalization. *J Biomed Inform, 47*, 1-10.
Doğan, R. I., & Lu, Z. (2012a). *An improved corpus of disease mentions in PubMed citations*. In Proceedings of the ACL 2012 Workshop on BioNLP, 91-99.
Doğan, R. I., & Lu, Z. (2012b). *An Inference Method for Disease Name Normalization*. In Proceedings of the AAAI 2012 Fall Symposium on Information Retrieval and Knowledge Discovery in Biomedical Text, 8-13.
Doğan, R. I., Murray, G. C., Névéol, A., & Lu, Z. (2009). Understanding PubMed user search behavior through log analysis. *Database (Oxford), 2009*, bap018.
Erhan, D., Bengio, Y., Courville, A., Manzagol, P.-A., Vincent, P., & Bengio, S. (2010). Why does unsupervised pre-training help deep learning? *J. Machine Learning Res., 11*, 625-660.
Finkel, J. R., Kleenman, A., & Manning, C. D. (2008). *Efficient, Feature-based, Conditional Random Field Parsing*. In Proceedings of the 46th Annual Meeting of the ACL, 959-967.
Gerner, M., Nenadic, G., & Bergman, C. M. (2010). LINNAEUS: a species name identification system for biomedical literature. *BMC Bioinformatics, 11*, 85.
Hirschman, L., Colosimo, M., Morgan, A., & Yeh, A. (2005). Overview of BioCreAtIvE task 1B: normalized gene lists. *BMC Bioinformatics, 6 Suppl 1*, S11.
Kang, N., Singh, B., Afzal, Z., van Mulligen, E. M., & Kors, J. A. (2012). Using rule-based natural language processing to improve disease normalization in biomedical text. *J. Am. Med. Inform. Assoc., 20*, 876-881.
Leaman, R., Doğan, R. I., & Lu, Z. (2013a). DNorm: Disease name normalization with pairwise learning-to-rank. *Bioinformatics, 29*(22), 2909-2917.
Leaman, R., & Gonzalez, G. (2008). BANNER: an executable survey of advances in biomedical named entity recognition. *Pac. Symp. Biocomput.*, 652-663.
Leaman, R., Khare, R., & Lu, Z. (2013b). *NCBI at 2013 ShARe/CLEF eHealth Shared Task: Disorder Normalization in Clinical Notes with DNorm*. In Working Notes of the Conference and Labs of the Evaluation Forum Valencia, Spain.
Lu, Z., Kao, H. Y., Wei, C. H., Huang, M., Liu, J., Kuo, C. J., et al. (2011). The gene normalization task in BioCreative III. *BMC Bioinformatics, 12 Suppl 8*, S2.
Manning, C., & Schütze, H. (1999). *Foundations of Statistical Natural Language Processing*: Massachusetts Institute of Technology.
Manning, C. D., Raghavan, P., & Schütze, H. (2008). *Introduction to Information Retrieval*: Cambridge University Press.
Mikolov, T., Yih, W.-t., & Zweig, G. (2013). *Linguistic Regularities in Continuous Space Word Representations*. In Proceedings of the 2013 Conference of the NAACL-HLT, 746-751.
Morgan, A. A., Lu, Z., Wang, X., Cohen, A. M., Fluck, J., Ruch, P., et al. (2008). Overview of BioCreative II gene normalization. *Genome Biol., 9 Suppl 2*, S3.
Névéol, A., Doğan, R. I., & Lu, Z. (2011). Semi-automatic semantic annotation of PubMed queries: a study on quality, efficiency, satisfaction. *J Biomed Inform, 44*(2), 310-318.
Névéol, A., Kim, W., Wilbur, W. J., & Lu, Z. (2009). *Exploring two biomedical text genres for disease recognition*. In Proceedings of the ACL 2009 BioNLP Workshop, 144-152.
Porter, M. F. (1980). An algorithm for suffix stripping. *Program, 14*, 130-137.
Suominen, H., Salanterä, S., Velupillai, S., Chapman, W., Savova, G., Elhadad, N., et al. (2013). Overview of the ShARe/CLEF eHealth Evaluation Lab 2013. In P. Forner, H. Müller, R. Paredes, P. Rosso & B. Stein (Eds.), *Information Access Evaluation. Multilinguality, Multimodality, and Visualization* (Vol. 8138, pp. 212-231): Springer Berlin Heidelberg.
Tibshirani, R. (1996). Regression shrinkage and selection via the Lasso. *Journal of the Royal Statistical Society Series B-Methodological, 58*(1), 267-288.
Tsuruoka, Y., McNaught, J., Tsujii, J., & Ananiadou, S. (2007). Learning string similarity measures for gene/protein name dictionary look-up using logistic regression. *Bioinformatics, 23*(20), 2768-2774.
Wei, C. H., Harris, B. R., Li, D., Berardini, T. Z., Huala, E., Kao, H. Y., et al. (2012a). Accelerating literature curation with text-mining tools: a case study of using PubTator to curate genes in PubMed abstracts. *Database (Oxford), 2012*, bas041.
Wei, C. H., Kao, H. Y., & Lu, Z. (2012b). SR4GN: a species recognition software tool for gene normalization. *PLoS One, 7*(6), e38460.
Wermter, J., Tomanek, K., & Hahn, U. (2009). High-performance gene name normalization with GeNo. *Bioinformatics, 25*(6), 815-821.

Decomposing Consumer Health Questions

Kirk Roberts, Halil Kilicoglu, Marcelo Fiszman, and Dina Demner-Fushman
National Library of Medicine
National Institutes of Health
Bethesda, MD 20894
robertske@nih.gov, {kilicogluh, fiszmanm, ddemner}@mail.nih.gov

Abstract

This paper presents a method for decomposing long, complex consumer health questions. Our approach largely decomposes questions using their syntactic structure, recognizing independent questions embedded in clauses, as well as coordinations and exemplifying phrases. Additionally, we identify elements specific to disease-related consumer health questions, such as the focus disease and background information. To achieve this, our approach combines rank-and-filter machine learning methods with rule-based methods. Our results demonstrate significant improvements over the heuristic methods typically employed for question decomposition that rely only on the syntactic parse tree.

1 Introduction

Natural language questions provide an intuitive method for consumers (non-experts) to query for health-related content. The most intuitive way for consumers to formulate written questions is the same way they write to other humans: multi-sentence, complex questions that contain background information and often more than one specific question. Consider the following:

- *Will Fabry disease affect a transplanted kidney? Previous to the transplant the disease was being managed with an enzyme supplement. Will this need to be continued? What cautions or additional treatments are required to manage the disease with a transplanted kidney?*

This complex question contains three question sentences and one background sentence. The focus (*Fabry disease*) is stated in the first question but is necessary for a full understanding of the other questions as well. The background sentence is necessary to understand the second question: the anaphor *this* must be resolved to *an enzyme treatment*, and the predicate *continue*'s implicit argument that must be re-constructed from the discourse (i.e., *continue after a kidney transplant*). The final question sentence uses a coordination to ask two separate questions (*cautions* and *additional treatments*). A decomposition of this complex question would then result in four questions:

1. *Will Fabry disease affect a transplanted kidney?*
2. *Will enzyme treatment for Fabry disease need to be continued after a kidney transplant?*
3. *What cautions are required to manage Fabry disease with a transplanted kidney?*
4. *What additional treatments are required to manage Fabry disease with a transplanted kidney?*

Each question above could be independently answered by a question answering (QA) system. While previous work has discussed methods for resolving co-reference and implicit arguments in consumer health questions (Kilicoglu et al., 2013), it does not address question decomposition.

In this work, we propose methods for automatically recognizing six annotation types useful for decomposing consumer health questions. These annotations distinguish between sentences that contain questions and background information. They also identify when a question sentence can be split in multiple independent questions, and

Proceedings of the 2014 Workshop on Biomedical Natural Language Processing (BioNLP 2014), pages 29–37,
Baltimore, Maryland USA, June 26-27 2014.

when they contain optional or coordinated information embedded within a question.

For each of these decomposition annotations, we propose a combination of machine learning (ML) and rule based methods. The ML methods largely take the form of a 3-step rank-and-filter approach, where candidates are generated, ranked by an ML classifier, then the top-ranked candidate is passed through a separate ML filtering classifier. We evaluate each of these methods on a set of 1,467 consumer health questions related to genetic and rare diseases.

2 Background

QA in the biomedical domain has been well-studied (Demner-Fushman and Lin, 2007; Cairns et al., 2011; Cao et al., 2011) as a means for retrieving medical information. This work has typically focused, however, on questions posed by medical professionals, and the methods proposed for question analysis generally assume a single, concise question. For example, Demner-Fushman and Abhyankar (2012) propose a method for extracting frames from queries for the purpose of cohort retrieval. Their method assumes syntactic dependencies exist between the necessary frame elements, and is thus not well-suited to handle long, multi-sentence questions. Similarly, Andersen et al. (2012) proposes a method for converting a concise question into a structured query. However, many medical questions require background information that is difficult to encode in a single question sentence. Instead, it is often more natural to ask multiple questions over several sentences, providing background information to give context to the questions. Yu and Cao (2008) use a ML method to recognize question types in professional health questions. Their method can identify more than one type per complex question. Without decomposing the full question into its sub-questions, however, the type cannot be associated with its specific span, or with other information specific to the sub-question. This other information can include answer types, question focus, and other answer constraints. By decomposing multi-sentence questions, these question-specific attributes can be extracted, and the discourse structure of the larger question can be better understood.

Question decomposition has been utilized before in open-domain QA approaches, but rarely evaluated on its own. Lacatusu et al. (2006) demonstrates how question decomposition can improve the performance of a multi-sentence summarization system. They perform what we refer to as *syntactic* question decomposition, where the syntactic structure of the question is used to identify sub-questions that can be answered in isolation. A second form of question decomposition is *semantic* decomposition, which can semantically break individual questions apart to answer them in stages. For instance, the question "*When did the third U.S. President die?*" can be semantically decomposed "*Who was the third U.S. President?*" and "*When did X die?*", where the answer to the first question is substituted into the second. Katz and Grau (2005) discusses this kind of decomposition using the syntactic structure, though it is not empirically validated. Hartrumpf (2008) proposes a decomposition method using only the deep semantic structure. Finally, Harabagiu et al. (2006) proposes a different type of question decomposition based on a random walk over similar questions extracted from a corpus. In our work, we focus on syntactic question decomposition. We demonstrate the importance of empirical evaluation of question decomposition, notably the pitfalls of heuristic approaches that rely entirely on the syntactic parse tree. Syntactic parsers trained on Treebank are particularly poor at both analyzing questions (Judge et al., 2006) and coordination boundaries (Hogan, 2007). Robust question decomposition methods, therefore, must be able to overcome many of these difficulties.

3 Consumer Health Question Decomposition

Our goal is to decompose multi-sentence, multi-faceted consumer health questions into concise questions coupled with important contextual information. To this end, we utilize a set of annotations that identify the decomposable elements and important contextual elements. A more detailed description of these annotations is provided in Roberts et al. (2014). The annotations are publicly available at our institution website[1]. Here, we briefly describe each annotation:

(1) BACKGROUND - a sentence indicating useful contextual information, but lacks a question.
(2) QUESTION - a sentence or clause that indicates an independent question.

[1] http://lhncbc.nlm.nih.gov/project/consumer-health-question-answering

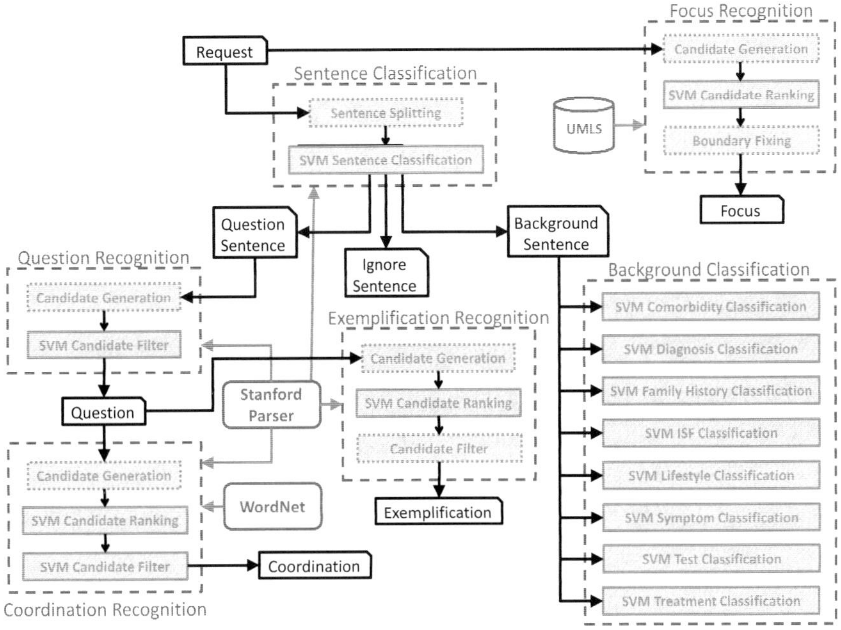

Figure 1: Question Decomposition Architecture. Modules with solid green lines indicate machine learning classifiers. Modules with dotted green lines indicate rule-based classifiers.

(3) COORDINATION - a phrase that spans a set of decomposable items.
(4) EXEMPLIFICATION - a phrase that spans an optional item.
(5) IGNORE - a sentence indicating nothing of value is present.
(6) FOCUS - an NP indicating the theme of the consumer health question.

Further explanations of each annotation are provided in Sections 4-9. To convert these annotations into separate, decomposed questions, a simple set of recursive rules is used. The rules enumerate all ways of including one conjunct from each COORDINATION as well as whether or not to include the phrase within an EXEMPLIFICATION. These rules must be applied recursively to handle overlapping annotations (e.g., a COORDINATION within an EXEMPLIFICATION). Our implementation is straight-forward and not discussed further in this paper. The BACKGROUND and FOCUS annotations do not play a direct role in this process, though they provide important contextual elements and are useful for co-reference, and are thus still considered part of the overall decomposition process.

It should also be noted that some questions are syntactically decomposable, but doing so alters their original meaning. Consider the following two question sentences:

- *Can this disease be cured or can we only treat the symptoms?*
- *Are males or females worse affected?*

While the first example contains two "*Can...*" questions and the second example contains the coordination "*males or females*", both questions are providing a choice between two alternatives and decomposing them would alter the semantic nature of the original question. In these cases, we do not consider the questions to be decomposable.

Data We use a set of consumer health questions collected from the Genetic and Rare Diseases Information Center (GARD), which maintains a website[2] with publicly available consumer-submitted questions and professionally-authored answers about genetic and rare diseases. We collected 1,467 consumer health questions, consisting of 4,115 sentences, 1,713 BACKGROUND sentences, 37 IGNORE sentences, 2,465 QUESTIONS, 367 COORDINATIONS, 53 EXEMPLIFICATIONS, and 1,513 FOCUS annotations. Questions with more than one FOCUS are generally concerned with the relation between diseases. Further information about the corpus and the annotation process can be found in Roberts et al. (2014).

System Architecture The architecture of our question decomposition method is illustrated in

[2]http://rarediseases.info.nih.gov/gard

Figure 1. To avoid confusion, in the rest of this paper we refer to a complex consumer health question simply as a *request*. Requests are sent to the independent FOCUS recognition module (Section 4), and then proceed through a pipeline that includes the classification of sentences (Section 5), the identification of separate QUESTIONs within a question sentence (Section 6), the recognition of COORDINATIONs (Section 7) and EXEMPLIFICATIONs (Section 8), and the sub-classification of BACKGROUND sentences (Section 9).

Experimental Setup The remainder of this paper describes the individual modules in Figure 1. For simplicity, we show results on the GARD data for each task in its corresponding section. In all cases, experiments are conducted using a 5-fold cross-validation on the GARD data. The cross-validation folds are organized at the request level so that no two items from the same request will be split between the training and testing data.

4 Identifying the Focal Disease

The FOCUS is the condition that disease-centered questions are centered around. Many other diseases may be mentioned, but the FOCUS is the disease of central concern. This is similar to the assumption made about a central disease in Medline abstracts (Demner-Fushman and Lin, 2007). Often the FOCUS is stated in the first sentence (typically a BACKGROUND) of the request while the questions are near the end. The questions cannot generally be answered outside the context of the FOCUS, however, so its identification is a critical part of decomposition. As shown in Figure 1, we use a 3-step process: (1) a high-recall method identifies potential FOCUS diseases in the data, (2) a support vector machine (SVM) ranks the FOCUS candidates, and (3) the highest-ranking candidate's boundary is modified with a set of rules to better match our annotation standard.

To identify candidates for the FOCUS, we use a lexicon constructed from UMLS (Lindberg et al., 1993). UMLS includes very generic terms, such as *disease* and *cancer*, that are too general to exactly match a FOCUS in our data. We allow these terms to be candidates so as to not miss any FOCUS that doesn't exactly match an entry in UMLS. When such a general term is selected as the top-ranked FOCUS, the rules described below are capable of expanding the term to the full disease name.

To rank candidates, we utilize an SVM (Fan et al., 2008) with a small number of feature types:

	E/R	P	R	F_1
1st UMLS *Disorder*	E	19.6	19.0	19.3
	R	28.2	27.4	27.8
SVM	E	56.4	54.7	55.6
	R	89.2	86.5	87.9
SVM + Rules	E	74.8	72.5	73.6
	R	89.5	86.8	88.1

Table 1: FOCUS recognition results. E = exact match; R = relaxed match.

- Unigrams. Identifies generic words such as *disease* and *syndrome* that indicate good FOCUS candidates, while also recognizing noisy UMLS terms that are often false positives.
- UMLS semantic group (McCray et al., 2001).
- UMLS semantic type.
- Sentence Offset. The FOCUS is typically in the first sentence, and is far more likely to be at the beginning of the request than the end.
- Lexicon Offset. The FOCUS is typically the first disease mentioned.

During training, the SVM considers any candidate that overlaps the gold FOCUS to be correct. This enables our approach to train on FOCUS examples that do not perfectly align with a UMLS concept. At test time, all candidates are classified, ranked by the classifier's confidence, and the top-ranked candidate is considered the FOCUS.

As mentioned above, there are differences between how a FOCUS is annotated in our data and how it is represented in the UMLS. We therefore use a series of heuristics to alter the boundary to a more usable FOCUS after it is chosen by the SVM. The rules are applied iteratively to widen the FOCUS boundary until it cannot be expanded any further. If a generic disease word is the only token in the FOCUS, we add the token to the left. Conversely, if the token on the right is a generic disease word, it is added as well. If the word to the left is capitalized, it is safe to assume it is part of the disease's name and so it is added as well. Finally, several rules recognize the various ways in which a disease sub-type might be specified (e.g., *Behcet's syndrome vascular type*, *type 2 diabetes*, *Charcot-Marie-Tooth disease type 2C*).

We evaluate FOCUS recognition with both an exact match, where the gold and automatic FOCUS boundaries must line up perfectly, and a relaxed match, which only requires a partial overlap. As a baseline, we compare our results against a fully rule-based system where the first UMLS *Disorder* term in the request is considered the FOCUS.

We also evaluate the effectiveness of our boundary altering rules by measuring performance without these rules. The results are shown in Table 1. The baseline method shows significant problems in precision and recall. It is not able to ignore noisy UMLS terms (e.g., *aim* is both a gene and a treatment). The SVM improves upon the rule-based method by over 50 points in F_1 for relaxed matching. Adding the boundary fixing rules has little effect on relaxed matching, but greatly improves exact matching: precision and recall are improved by 18.4 and 17.8 points, respectively.

5 Classifying Sentences

Before precise question boundaries can be recognized, we first identify sentences that contain QUESTIONS, as distinguished from BACKGROUND and IGNORE sentences. It should be noted that many of the question sentences in our data are not typical wh-word questions. About 20% of the questions in our data end in a period. For instance:

- *Please tell me more about this condition.*
- *I was wondering if you could let me know where I can find more information on this topic.*
- *I would like to get in contact with other families that have this illness.*

We consider a sentence to be a question if it contains any information request, explicit or implicit.

After sentence splitting, we identify sentences using a multi-class SVM with three feature types:

- Unigrams with parts-of-speech (POS). Reduces unigram ambiguities, such as *what-WP* (a pronoun, indicative of a question) versus *what-WDT* (a determiner, not indicative).
- Bigrams.
- Parse tree tags. All Treebank tags from the syntactic parse tree. Captures syntactic question clues such as the phrase tags *SQ* (question sentence) and *WHNP* (wh-word noun phrase).

The SVM classifier performs at 97.8%. For comparison, an SVM with only unigram features performs at 97.2%. While the unigram model does a good job classifying sentences, suggesting this is a very easy task, the improved feature set reduces the number of errors by 20%.

6 Identifying Questions

QUESTION recognition is the task of identifying when a conjunction like *and* joins two independent questions into a single sentence:

- [*What causes the condition*]$_{\text{QUESTION}}$ [*and what treatment is available?*]$_{\text{QUESTION}}$
- [*What is this disease*]$_{\text{QUESTION}}$ [*and what steps can I take to protect my daughter?*]$_{\text{QUESTION}}$

We consider the identification of separate QUESTIONS within a single sentence to be a different task from COORDINATION recognition, which finds phrases whose conjuncts can be treated independently. Linguistically, these tasks are quite similar, but the distinction lies in whether the right-conjunct syntactically depends on anything to its left. For instance:

- *I would like to learn* [*more about this condition* **and** *what the prognosis is for a baby born with it*]$_{\text{COORDINATION}}$.

Here, the right-conjunct starts with a question stem (*what*), but is not a complete, grammatical question on its own. Alternatively, this could be re-formed into two separate QUESTIONS:

- [*I would like to learn more about this condition,*]$_{\text{QUESTION}}$ [**and** *what is the prognosis is for a baby born with it.*]$_{\text{QUESTION}}$

We make this distinction because the QUESTION recognition task requires one fewer step since the boundaries extend to the entire sentence, preventing error propagation from an input module. Further, the features that differentiate our QUESTION and COORDINATION annotations are different.

The two-step process for recognizing QUESTIONS includes: (1) a high-recall candidate generator, and (2) an SVM to eliminate candidates that are not separate QUESTIONS. The candidates for QUESTION recognition are simply all the ways a sentence can be split by the conjunctions *and*, *or*, *as well as*, and the forward slash ("/"). In our data, this candidate generation process has a recall of 98.6, as a few examples were missed where candidates were not separated by one of the above conjunctions.

To filter candidates, we use an SVM with three features types:

- The conjunction separating the QUESTIONS.
- Unigrams in the left-conjunct. Identifies when the left-conjunct is not a QUESTION, or when a question is part of a COORDINATION.
- The right-conjunct's parse tree tag. Recognizes when the right-conjunct is an independent clause that may safely be split.

	P	R	F_1
QUESTION split recognition			
Baseline	24.7	82.4	38.0
SVM	67.7	64.7	66.2
Overall QUESTION recognition			
Baseline	87.3	92.8	90.0
SVM	97.7	97.4	97.5

Table 2: QUESTION recognition results.

For evaluation, we measure both the F_1 score for correct candidates, and the overall F_1 for all QUESTION annotations (i.e., all QUESTION sentences). We also evaluate a baseline method that utilizes the parse tree to recognize separate QUESTIONs by splitting sentences where a conjunction separates independent clauses. The results are shown in Table 2. The baseline method has good recall for recognizing where a sentence should be split into multiple QUESTIONs, but it lacks precision. This is largely because it is unable to differentiate clausal COORDINATIONs such as the above example, as well as when the left-conjunct is not actually a separate question. For instance:

- *Our grandson was diagnosed recently with this disease* **and** *I am wondering if you could send me information on it.*

The SVM-based method can overcome this problem by looking at the words in the left-conjunct. Both methods, however, fail to recognize when two independent question clauses are asking the same question but providing alternative answers:

- *Will this condition be with him throughout his life,* **or** *is it possible that it will clear up?*

While there are methods for handling this issue for COORDINATION recognition, addressed below, recognizing non-splittable QUESTIONs requires far deeper semantic understanding which we leave to future work.

7 Identifying Coordinations

COORDINATION recognition is the task of identifying when a conjunction joins phrases within a QUESTION that can in be separate questions:

- *How can I learn more about [treatments* **and** *clinical trials]*$_{\text{COORDINATION}}$*?*
- *Are [muscle twitching, muscle cramps, and muscle pain]*$_{\text{COORDINATION}}$ *effects of having silicosis?*

Unlike QUESTION recognition, the boundaries of a COORDINATION need to be determined as well as whether the conjuncts can semantically be split into separate questions. We thus use a three-step process for recognizing COORDINATIONs: (1) a high-recall candidate generator, (2) an SVM to rank all the candidates for a given conjunction, and (3) an SVM to filter out top-ranked candidates.

Candidate generation begins with the identification of valid conjunctions within a QUESTION annotation. We use the same four conjunctions as in QUESTION recognition: *and*, *or*, *as well as*, and the forward slash. For each of these, all possible left and right boundaries are generated, so in a QUESTION with 4 tokens on either side of the conjunction, there would be 16 candidates. Additionally, two adjectives separated by a comma and immediately followed by a noun are considered a candidate (e.g., "*a [safe, permanent]*$_{\text{COORDINATION}}$ *treatment*"). In our data, this candidate generation process has a recall of 98.9, as a few instances exist in which a conjunction is not used, such as:

- *I am looking for any information you have about heavy metal toxicity, [treatment, outcomes]*$_{\text{EXEMPLIFICATION+COORDINATION}}$.

To rank candidates, we use an SVM with the following feature types:

- If the left-conjunct is congruent with the highest node in the syntactic parse tree whose right-most leaf is also the right-most token in the left-conjunct. Essentially, this is equivalent to saying whether or not the syntactic parser agrees with the left-conjunct's boundary.
- The equivalent heuristic for the right-conjunct.
- If a noun is in *both*, just the *left* conjunct, just the *right* conjunct, or *neither* conjunct.
- The Levenshtein distance between the POS tag sequences for the left- and right-conjuncts.

The first two features encode the information a rule-based method would use if it relied entirely on the syntactic parse tree. The remaining features help the classifier overcome cases where the parser may be wrong.

At training time, all candidates for a given conjunction are generated and only the candidate that matches the gold COORDINATION is considered a positive example. Additionally, we annotated the boundaries for negative COORDINATIONs (i.e., syntactic coordinations that do not fit our annotation standard). There were 203 such instances in the GARD data. These are considered gold COORDINATIONs for boundary ranking only.

To filter the top-ranked candidates, we use an SVM with several feature types:

	E/R	P	R	F_1
Baseline	E	28.1	36.5	31.8
	R	62.9	75.8	68.7
Rank + Filter	E	38.2	34.8	36.4
	R	78.5	69.0	73.5

Table 3: COORDINATION recognition results. E = exact match; R = relaxed match.

- The conjunction.
- Unigrams in the left-conjunct.
- POS of the first word in both conjuncts. COORDINATIONS often have the same first POS in both conjuncts.
- The word immediately before the candidate. E.g., *between* is a good negative indicator.
- Unigrams in the question but not the candidate.
- If the candidate takes up almost the entire question (all but 3 tokens). Typically, COORDINATIONS are much smaller than the full question.
- If more than one conjunction is in the candidate.
- If a word in the left-conjunct has an antonym in the right conjunct. Antonyms are recognized via WordNet (Fellbaum, 1998).

At training time, the positive examples are drawn from the annotated COORDINATIONS, while the negative examples are drawn from the 203 non-gold annotations mentioned above.

In addition to evaluating this method, we evaluate a baseline method that relies entirely on the syntactic parse to identify COORDINATION boundaries without filtering. The results are shown in Table 3. The rank-and-filter approach shows significant gains over the rule-based method in precision and F_1. As can be seen in the difference between exact and relaxed matching, most of the loss for both the baseline and ML methods come in boundary detection. Most methods overly rely upon the syntactic parser, which performs poorly both on questions and coordinations. The ML method, though, is sometimes able to overcome this problem.

8 Identifying Exemplifications

EXEMPLIFICATION recognition is the task of identifying when a phrase provides an optional, exemplifying example with a more specific type of information than that asked by the rest of the question. For instance, the following contains both an EXEMPLIFICATION and a COORDINATION:

- *Is there anything out there that can help him [such as [medications or alternative therapies]$_{\text{COORDINATION}}$]$_{\text{EXEMPLIFICATION}}$?*

We could consider this to denote 3 questions:

- *Is there anything out there that can help him?*
- *Is there anything out there that can help him such as medications?*
- *Is there anything out there that can help him such as alternative therapies?*

In the latter two questions, we consider the phrase *such as* to now denote a mandatory constraint on the answer to each question, whereas in the original question it would be considered optional.

EXEMPLIFICATION recognition is similar to COORDINATION recognition, and its three-step process is thus similar as well: (1) a high-recall candidate generator, (2) an SVM to rank all the candidates for a given trigger phrase, and (3) a set of rules to filter out top-ranked candidates.

Candidate generation begins with the identification of valid trigger words and phrases. These include: *especially*, *including*, *particularly*, *specifically*, and *such as*. For each of these, all possible right boundaries are generated, thus EXEMPLIFICATIONS have far fewer candidates than COORDINATIONS. Additionally, all phrases within parentheses are added as EXEMPLIFICATIONS. In our data, this candidate generation process has a recall of 98.1, missing instances without a trigger (see the example also missed by COORDINATION candidate generation in Section 7).

To rank candidates, we use an SVM with the following feature types:

- If the right-conjunct is the highest parse node as defined in the COORDINATION boundary feature.
- If a dependency relation crosses from the right-conjunct to any word outside the candidate.
- POS of the word after the candidate.

As with COORDINATIONS, we annotated boundaries for negative EXEMPLIFICATIONS matching the trigger words and used them as positive examples for boundary ranking.

To filter the top-ranked candidates, we use two simple rules. First, EXEMPLIFICATIONS within parentheses are filtered if they are acronyms or acronym expansions. Second, cases such as the below example are removed by looking at the words before the candidate:

- *I am* **particularly** *interested in learning more about genetic testing for the syndrome.*

In addition to evaluating this method, we evaluate a baseline method that relies entirely on the

	E/R	P	R	F_1
Baseline	E	28.9	62.3	39.5
	R	39.5	84.9	53.9
Rank + Filter	E	60.8	58.5	59.6
	R	80.4	77.4	78.8

Table 4: EXEMPLIFICATION recognition results. E = exact match; R = relaxed match.

syntactic parser to identify EXEMPLIFICATION boundaries and performs no filtering. The results are shown in Table 4. The rank-and-filter approach shows significant gains over the rule-based method in precision and F_1, more than doubling precision for both exact and relaxed matching. There is still a drop in performance when going from relaxed to exact matching, again largely due to the reliance on the syntactic parser.

9 Classifying Background Information

BACKGROUND sentences contain contextual information, such as whether or not a patient has been diagnosed with the focal disease or what symptoms they are experiencing. This information was annotated at the sentence level, partly because of annotation convenience, but also because phrase boundaries are not always clear for medical concepts (Hahn et al., 2012; Forbush et al., 2013).

A difficult factor in this task, and especially on the GARD dataset, is that consumers are not always asking about a disease for themselves. Instead, often they ask on behalf of another individual, often a family member. The BACKGROUND types are thus annotated based on the person of interest, who we refer to as the *patient* (in the linguistic sense). For instance, if a mother has a disease but is asking about her son (e.g., asking about the probability of her son inheriting the disease), that sentence would be a FAMILY_HISTORY, as opposed to a DIAGNOSIS sentence.

The GARD corpus is annotated with eight BACKGROUND types:

- COMORBIDITY
- DIAGNOSIS
- FAMILY_HISTORY
- ISF (information search failure)
- LIFESTYLE
- SYMPTOM
- TEST
- TREATMENT

ISF sentences indicate previous attempts to find the requested information have failed, and are a good signal to the QA system to enable more in-depth search strategies. LIFESTYLE sentences describe the patient's life habits (e.g., smoking, exercise). Currently, the automatic identification of

Type	P	R	F_1	# Anns
COMORBIDITY	0.0	0.0	0.0	23
DIAGNOSIS	80.8	80.3	80.5	690
FAMILY_HISTORY	67.4	38.4	48.9	151
ISF	75.0	65.9	70.1	41
LIFESTYLE	0.0	0.0	0.0	13
SYMPTOM	76.6	48.1	59.1	320
TEST	37.5	4.9	8.7	61
TREATMENT	87.3	35.0	50.0	137
Overall: Micro-F_1: 67.3 Macro-F_1: 39.7				

Table 5: BACKGROUND results.

BACKGROUND types has not been a major focus of our effort as no handling exists for it within our QA system. We report a baseline method and results here to provide some insight into the difficulty of the task.

BACKGROUND types are a multi-labeling problem, so we use eight binary classifiers, one for each type. Each classifier utilizes only unigram and bigram features. The results for the models are shown in Table 5. COMORBIDITY and LIFESTYLE are too rare in the data (23 and 13 instances, respectively) for the classifier to identify. DIAGNOSIS questions are identified fairly well because this is the most common type (690 instances) and because of the constrained vocabulary for expressing a diagnosis. The performance of the rest of the types is largely proportional to the number of instances in the data, though ISF performs quite well given only 41 instances.

10 Conclusion

We have presented a method for decomposing consumer health questions by recognizing six annotation types. Some of these types are general enough to use in open-domain question decomposition (BACKGROUND, IGNORE, QUESTION, COORDINATION, EXEMPLIFICATION), while others are targeted specifically at consumer health questions (FOCUS and the BACKGROUND subtypes). We demonstrate that ML methods can improve upon heuristic methods relying on the syntactic parse tree, though parse errors are often difficult to overcome. Since significant improvements in performance would likely require major advances in open-domain syntactic parsing, we instead envision further integration of the key tasks in consumer health question analysis: (1) integration of co-reference and implicit argument information, (2) improved identification of BACKGROUND types, and (3) identification of discourse relations within questions to further leverage question decomposition.

Acknowledgements

This work was supported by the intramural research program at the U.S. National Library of Medicine, National Institutes of Health. We would additionally like to thank Stephanie M. Morrison and Janine Lewis for their help accessing the GARD data.

References

Ulrich Andersen, Anna Braasch, Lina Henriksen, Csaba Huszka, Anders Johannsen, Lars Kayser, Bente Maegaard, Ole Norgaard, Stefan Schulz, and Jürgen Wedekind. 2012. Creation and use of Language Resources in a Question-Answering eHealth System. In *Proceedings of the Eighth International Conference on Language Resources and Evaluation*, pages 2536–2542.

Brian L. Cairns, Rodney D. Nielsen, James J. Masanz, James H. Martin, Martha S. Palmer, Wayne H. Ward, and Guergana K. Savova. 2011. The MiPACQ Clinical Question Answering System. In *Proceedings of the AMIA Annual Symposium*, pages 171–180.

YongGang Cao, Feifan Liu, Pippa Simpson, Lamont Antieau, Andrew Bennett, James J. Cimino, John Ely, and Hong Yu. 2011. AskHERMES: An online question answering system for complex clinical questions. *Journal of Biomedical Informatics*, 44:277–288.

Dina Demner-Fushman and Swapna Abhyankar. 2012. Syntactic-Semantic Frames for Clinical Cohort Identification Queries. In *Data Integration in the Life Sciences*, volume 7348 of *Lecture Notes in Computer Science*, pages 100–112.

Dina Demner-Fushman and Jimmy Lin. 2007. Answering Clinical Questions with Knowledge-Based and Statistical Techniques. *Computational Linguistics*, 33(1).

Rong-En Fan, Kai-Wei Chang, Cho-Jui Hsieh, Xiang-Rui Wang, and Chih-Jen Lin. 2008. LIBLINEAR: A Library for Large Linear Classification. *Journal of Machine Learning Research*, 9:1871–1874.

Christiane Fellbaum. 1998. *WordNet: An Electronic Lexical Database*. MIT Press.

Tyler B. Forbush, Adi V. Gundlapalli, Miland N. Palmer, Shuying Shen, Brett R. South, Guy Divita, Marjorie Carter, Andrew Redd, Jorie M. Butler, and Matthew Samore. 2013. "Sitting on Pins and Needles": Characterization of Symptom Descriptions in Clinical Notes. In *AMIA Summit on Clinical Research Informatics*, pages 67–71.

Udo Hahn, Elena Beisswanger, Ekaterina Buyko, Erik Faessler, Jenny Traumüller, Susann Schröder, and Kerstin Hornbostel. 2012. Iterative Refinement and Quality Checking of Annotation Guidelines – How to Deal Effectively with Semantically Sloppy Named Entity Types, such as Pathological Phenomena. In *Proceedings of the Eighth International Conference on Language Resources and Evaluation*, pages 3881–3885.

Sanda Harabagiu, Finley Lacatusu, and Andrew Hickl. 2006. Answer Complex Questions with Random Walk Models. In *Proceedings of the 29th Annual ACM SIGIR Conference on Research and Development in Information Retrieval*, pages 220–227.

Sven Hartrumpf. 2008. Semantic Decomposition for Question Answering. In *Proceedings on the 18th European Conference on Artificial Intelligence*, pages 313–317.

Dierdre Hogan. 2007. Coordinate Noun Phrase Disambiguation in a Generative Parsing Model. In *Proceedings of the 45th Annual Meeting of the Association for Computational Linguistics*, pages 680–687.

John Judge, Aoife Cahill, and Josef van Genabith. 2006. QuestionBank: Creating a Corpus of Parse-Annotated Questions. In *Proceedings of the 21st International Conference on Computational Linguistics and 44th Annual Meeting of the Association for Computational Linguistics*, pages 497–504.

Yarden Katz and Bernardo C. Grau. 2005. Representing Qualitative Spatial Information in OWL-DL. *Proceedings of OWL: Experiences and Directions*.

Halil Kilicoglu, Marcelo Fiszman, and Dina Demner-Fushman. 2013. Interpreting Consumer Health Questions: The Role of Anaphora and Ellipsis. In *Proceedings of the 2013 BioNLP Workshop*, pages 54–62.

Finley Lacatusu, Andrew Hickl, and Sanda Harabagiu. 2006. Impact of Question Decomposition on the Quality of Answer Summaries. In *Proceedings of LREC*, pages 1147–1152.

Donald A.B. Lindberg, Betsy L. Humphreys, and Alexa T. McCray. 1993. The Unified Medical Language System. *Methods of Information in Medicine*, 32(4):281–291.

Alexa T McCray, Anita Burgun, and Olivier Bodenreider. 2001. Aggregating UMLS Semantic Types for Reducing Conceptual Complexity. In *Studies in Health Technology and Informatics (MEDINFO)*, volume 84(1), pages 216–220.

Kirk Roberts, Kate Masterton, Marcelo Fiszman, Halil Kilicoglu, and Dina Demner-Fushman. 2014. Annotating Question Decomposition on Complex Medical Questions. In *Proceedings of LREC*.

Hong Yu and YongGang Cao. 2008. Automatically Extracting Information Needs from Ad Hoc Clinical Questions. In *Proceedings of the AMIA Annual Symposium*.

Detecting Health Related Discussions in Everyday Telephone Conversations for Studying Medical Events in the Lives of Older Adults

Golnar Sheikhshab, Izhak Shafran, Jeffrey Kaye
Oregon Health & Science University
sheikhsh,shafrani,kaye@ohsu.edu

Abstract

We apply semi-supervised topic modeling techniques to detect health-related discussions in everyday telephone conversations, which has applications in large-scale epidemiological studies and for clinical interventions for older adults. The privacy requirements associated with utilizing everyday telephone conversations preclude manual annotations; hence, we explore semi-supervised methods in this task. We adopt a semi-supervised version of Latent Dirichlet Allocation (LDA) to guide the learning process. Within this framework, we investigate a strategy to discard irrelevant words in the topic distribution and demonstrate that this strategy improves the average F-score on the in-domain task and an out-of-domain task (Fisher corpus). Our results show that the increase in discussion of health related conversations is statistically associated with actual medical events obtained through weekly self-reports.

1 Introduction

There has been considerable interest in understanding, promoting, and monitoring healthy lifestyles among older adults while minimizing the frequency of clinical visits. Longitudinal studies on large cohorts are necessary, for example, to understand the association between social networks, depression, dementia, and general health. In this context, detecting discussions of health are important as indicators of under-reported health events in daily lives as well as for studying healthy social support networks. The detection of medical events such as higher levels of pain or discomfort may also be useful in providing timely clinical intervention for managing chronic illness and thus promoting healthy independent living among older adults.

Motivated by this larger goal, we develop and investigate techniques for identifying conversations containing any health related discussion. We are interested in detecting discussions about medication with doctors, as well as conversations with others, where among all different topics being discussed, subjects may also be complaining about pain or changes in health status.

The privacy concerns of recording and analyzing everyday telephone conversation prevents us from manually transcribing and annotating conversations. So, we automatically transcribe the conversations using an automatic speech recognition system and look-up the telephone number corresponding to each conversation as a heuristic means of deriving labels. This technique is suitable for labeling a small subset of the conversations that are only sufficient for developing semi-supervised algorithms and for evaluating the methods for analysis.

Before delving into our approach, we discuss a few relevant and related studies in Section 2 and describe our unique naturalistic corpus in Section 3. Given the restrictive nature of our labeled in-domain data set, we are interested in a classifier that generalizes to the unlabeled data. We evaluate the generalizability of the classifiers using an out-of-domain corpus. We adopt a semi-supervised topic modeling approach to address our task, and develop an iterative feature selection method to improve our classifier, as described in Section 4. We evaluate the efficacy of our approach empirically, on the in-domain as well as an out-of-domain corpus, and report results in Section 5.

2 Related Work

The task of identifying conversations where health is mentioned differs from many other tasks in topic

Proceedings of the 2014 Workshop on Biomedical Natural Language Processing (BioNLP 2014), pages 38–44,
Baltimore, Maryland USA, June 26-27 2014.

modeling because in this task we are interested in one particular topic. A similar study is the work of Prier and colleagues (Prier et al., 2011). They use a set of predefined seed words as queries to gather tweets related to tobacco or marijuana usage, and then use LDA to discover related subtopics. Thus, their method is sensitive to the seed words chosen.

One way to reduce the sensitivity to the manually specified seed words is to expand the set using WordNet. Researchers have investigated this approach in sentiment analysis (Kim and Hovy, 2004; Yu and Hatzivassiloglou, 2003). However, when expanding the seed word set using WordNet, we need to be careful to avoid antonyms and words that have high degree of linkage with many words in the vocabulary. Furthermore, we can not apply such an approach for languages with poor resources, where manually curated knowledge is unavailable. The other drawback of this approach is that we can not use characteristics of the end task, in our case health-related conversation retrieval, to select the words. As an alternative method, Han and colleagues developed an interactive system where users selected the most relevant words from a set, proposed by an automated system (Han et al., 2009).

Another idea for expanding the seed words is using the statistical information. Among statistical methods, the simplest approach is to compute pairwise co-occurrence with the seed words. Li and Yamanishi ranked the words co-occurring with the seed words according to information theoretic costs, and used the highest ranked words as the expanded set (Li and Yamanishi, 2003). This idea can be more effective when the co-occurrence is performed over subsets instead, as in Hisamitsu and Niwa's work (Hisamitsu and Niwa, 2001). However, it is computationally expensive to search over subsets of words. Depending on the language and task, heuristics might be applicable. An example of this kind of approach is Zagibalov and Carroll's work on sentiment analysis in Chinese (Zagibalov and Carroll, 2008).

Alternatively, we can treat the task of identifying words associated with seed words as a clustering problem with the intuition that the seed words are in the same cluster. An effective strategy to cluster words into topics, is Latent Dirichlet Allocation (LDA) (Blei et al., 2003) . However, LDA is an unsupervised algorithm and the clustered topics are not guaranteed to include the topic of interest. The Seeded LDA, a variant of LDA, attempts to address this problem by incorporating the seed words as priors over the topics (Jagarlamudi et al., 2012). However, the estimation procedure is more complicated. Alternatively, in Topic LDA (TLDA), a clever extension to LDA, Andrzejewski and Zhu address this problem by fixing the membership of the words to valid topics (Andrzejewski and Zhu, 2009). When the focus is on detecting just one topic, as in our task, we can expand the seed words more selectively using the small set of labeled data and that is the approach adopted in this paper.

3 Data

One interesting aspect of our study is the uniqueness of our corpus, which is both naturalistic and exhaustive. We recorded about 41,000 land-line everyday telephone conversations from 56 volunteers, 65 years or older, over a period of approximately 6 to 12 months. Since these everyday telephone conversations are private conversations, and might include private information such as names, telephone numbers, or banking information, we assured the subjects that no one would listen to the recorded conversations. Thus, we couldn't manually transcribe the conversations; instead, we used an Automatic Speech Recognition (ASR) system that we describe here.

Automatic Speech Recognition System Conversations in our corpus were automatically transcribed using an ASR system, which is structured after IBM's conversation telephony system (Soltau et al., 2005). The acoustic models were trained on about 2000 hours of telephone speech from Switchboard and Fisher corpora (Godfrey et al., 1992). The system has a vocabulary of 47K and uses a trigram language model with about 10M n-grams, estimated from a mix of transcripts and web-harvested data. Decoding is performed in three stages using speaker-independent models, vocal-tract normalized models and speaker-adapted models. The three sets of models are similar in complexity with 4000 clustered pentaphone states and 150K Gaussians with diagonal covariances. Our system does not include discriminative training and performs at a word error rate of about 24% on NIST RT Dev04 which is comparable to state of the art performance for such systems. We are unable to measure the performance of this recognizer on our corpus due to the stringent privacy requirements mentioned earlier. Since both corpora are conversational telephone speech and the training data contains large number of conversations (2000 hours), we expect the performance of our recognizer to be relatively close to results on NIST benchmark.

Heuristically labeling a small subset of conversations For training and evaluation purposes, we need a labeled set of conversations; that is, a set of conversations where we know whether or not they contain health-related discussions. Since the privacy concerns do not allow for manually labeling the conversations, we used reverse look-up service in `www.whitepages.com`. We sent the phone number corresponding to each conversation (when available) to this website to obtain information about the other end of the conversation. Based of the information we got back from this website, we labeled a small subset of the conversations which fell into unambiguous business categories. For example, we labeled the calls to "hospital" and "pharmacy" as health-related, and those to "car repair" and "real estate" as non-health-related.

The limitations of the labeled set The labeled set we obtained is small and restricted in type of conversations. Since phone numbers are not available for many of the conversations we recorded, and also because `www.whitepages.com` does not return unambiguous information for many of available phone numbers, we managed to label only 681 conversations – 275 health-related and 406 non-health-related. This labeled set has another limitation: it contains conversations to business numbers only. In reality however, we are interested in the much larger set of conversations between friends, relatives, and other members of subjects' social support network. Thus, the generalizability of the classifier we train is very important.

Fisher Corpus To explicitly test the generalizability of our classifier, we use a second evaluation set from Fisher corpus (Cieri et al., 2004). Fisher corpus contains telephone conversations with pre-assigned topics. There are 40 topics and only one of them, illness, is health-related. We identified 338 conversations on illness, and sampled 702 conversations from the other 39 non-health topics. Since we do not train on Fisher corpus, we call it the out-of-domain task to apply our method on Fisher corpus; as opposed to the in-domain task which is to apply our method on the everyday telephone conversations.

Extra information on subjects' health In the everyday telephone conversations corpus, we also have access to the subjects' weekly self-reports on their medical status during the week indicating medical events such as injury or going to emergency room. We will use these pieces of information to relate the health-related conversations to actual medical events in the subjects' lives.

4 Method

4.1 Overview

As we explained in Section 3, we can label a small set of conversations in the everyday telephone conversations corpus as health-related vs. non-health related. Using this labeled set we can train a support vector machine (SVM) to classify the conversations. In absence of feature selection, the conversations are represented by a vector of tf-idf scores for every word in the vocabulary where tf-idf is a score for measuring the importance of a word in one document of a corpus. As we see in Section 5, such a classifier doesn't generalize to the out-of-domain Fisher task (*i.e.* when we test the classifier on Fisher data set, we do not get good precision and recalls). Generalizability is important in our case, especially because the data we use for training is limited in number and the nature of conversations.

One way to improve generalization is to perform feature selection. That is, instead of using tf-idf scores for the whole vocabulary, we would like to rely only on features relevant to detecting the health topic. We propose a new way for feature selection for retrieving documents containing information about a specific topic when there is only a limited set of labeled documents available. The idea is to pick a few words highly related to the topic of interest as seed words and to use TLDA (Andrzejewski and Zhu, 2009) to force those seed words into one (for example, the first) topic. In our task, the topic of interest is health. So, we choose *doctor, medicine,* and *pain* – often used while discussing health – as our seed words. Topics in LDA based methods such as TLDA are usually represented using the n most probable words; where n is an arbitrary number. So, the first candidate sets for expanding our seed words are the sets of 50 most probable words in the topic of health in different runs of TLDA. As our experiments reveal, these candidate sets contain many words that are unrelated to health . To solve this problem, we use the small labeled set of conversations to filter out the unrelated words.

Figure 1 shows the proposed iterative algorithm. The algorithm starts with initializing the seed words to *doctor, medicine, and pain*. Then, in each iteration, TLDA performs semi-supervised topic modeling and returns the 50 most probable candidate words in the health topic. We select a subset of these candidate words which, if added to the seed words, would maximize the average of precision and recall on the train set for a simple classifier. This simple classifier marks a conversion as health related if, and only if, it contains at least one of the seed words. The algorithm terminates when the subset selection is unable to add a new word contributing to the average of precision and recall. The tf-idf vector for the expanded set represents the conversations in the classification process.

It is worth mentioning that we train TLDA using all 41000 unlabeled conversations, and chose the number of topics, K, to be 20.

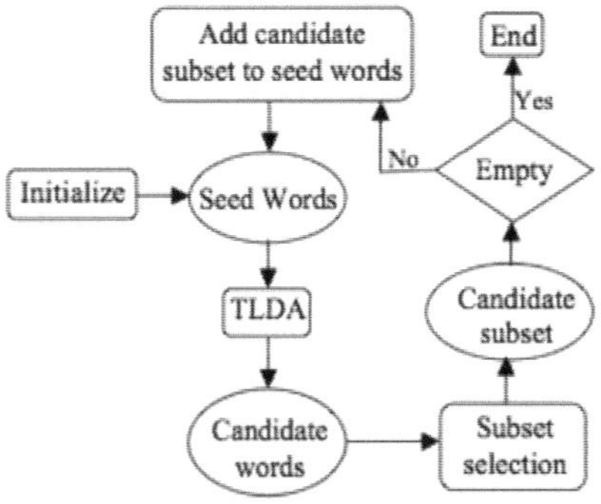

Figure 1: *Expanding the set of seed words:* in each iteration, the current seed words are forced into the topic of health to guide TLDA towards finding more health related words. The candidate set consists of the 50 most probable words of the topic of health in TLDA. We investigate the gain of adding each word of the candidate set to the seed words by temporarily adding it to the seed words and looking at the average of precision and recall on the training set for a classifier that classifies a conversation as health-related if and only if it contains at least one of the seed words. We select the words that maximize this objective and add them to the seed words until no other words contributes to the average precision and recall.

5 Experiments

In all of our experiments, we trained SVM classifiers, with different features, to detect the conversations on health using the popular libSVM (Chang and Lin, 2011) implementation. We chose the parameters of the SVM using a 30-fold cross-validated (CV) grid search over the training data. We also used a 4-fold cross validation over the labeled set of conversations to maximize the use of the relatively small labeled set. That is, we trained the feature selection algorithm on 3-folds and tested the resulting SVM tested on the fourth. In in-domain task we always report the average performance across the folds.

Table 1 shows the results of our experiments using different input features. We report on recall, precision and F-measure in in-domain and out-of-domain (Fisher) task as well as on average F-measure of the two. The justification for considering the average F-measure is that we want our algorithm to work well on both in-domain corpus and Fisher corpus since we need to make sure that our classifier is generalizable (i.e. it works well on Fisher) and it works well on the private and natural telephone conversations (i.e. the ones similar to the in-domain corpus)

When using the full vocabulary, the in-domain performance (the performance on the everyday telephone conversations data) is relatively good with 75.1% recall and 83.5% precision. But the out-of-domain recall (recall on the Fisher data set) is considerably low at 2.8%. Ideally, we want a classifier that performs well in both domains. Rows 2 to 5 can be seen as steps to get to such a classifier.

The second row shows the performance of the other extreme end of feature selection: the features include the manually chosen words *doctor, medicine, and pain* only. While this leads to very good out-of-domain performance, the in-domain recall has dropped considerably. We trained TLDA 30 times, and selected the 50 most probable words in the health topic. The third row in Table 1 shows the average performance of SVM when using the tf-idf of these sets of words as the feature vector on in-domain and out-of-domain tasks. Using the 50 most probable words in health topic significantly improves average F-score (71%) across

	Recall		Precision		F-measure		
Feature Words	In-Domain	Fisher	In-Domain	Fisher	In-Domain	Fisher	Average
Full vocabulary (no feature selection)	75.2	2.8	83.5	91.1	79.1	5.4	42.3
Initial words (*doctor, medicine, pain*)	45.1	69.2	94.8	94.5	61.1	79.9	70.5
50 most probable words in *health* (average over 30 runs)	58.4	57.4	86.3	97.5	69.7	72.3	71.0
Words selected by our method (average over 30 runs)	56.1	66.5	91.0	95.5	69.4	78.4	73.9
Union of all selected words (across 30 runs)	67.7	69.4	87.8	95.1	76.5	80.2	78.3

Table 1: Performance of SVM classifiers using different feature selection methods. The In-Domain task involves the everyday telephone conversations corpus. We call Fisher corpus out of domain, because no example of this corpus was used in training.

both tasks over using the full vocabulary (42.3%) but it is clear that this is only due to improvement in out-of-domain task. Table 2 shows one set of the 50 most probable words in health topic,the result of one run of TLDA. Evidently, these words contain many irrelevant words. This is the motivation for our iterative algorithm.

Next, we evaluate the performance of our iterative algorithm. The fourth row in Table 1 shows the average performance of SVM using expanded seed words that our algorithm suggested in 30 runs. Our algorithm improves the average F-score by 3% comparing to the standard TLDA. This is due to a 5% improvement in out-of-domain task as opposed to a 0.3% performance decrease in in-domain task.

Since our algorithm has a probabilistic topic modeling component (*i.e.* TLDA), different runs lead to different sets of expanded seed words. We extract a union of all the words chosen over 30 runs and evaluate the performance of SVM using this union set. This improves the performance of our method further to achieve the best average F-score of 78.3%, which is an 85% improvement over using the SVM with full vocabulary. It is important to notice that the in-domain performance is still lower than the full-vocabulary baseline by less than 3% while the out-of-domain performance is the best obtained. Once again, we are more interested in the average F-measure because we need our algorithm to generalize well (work well on out-of-domain corpus) and to work well on natural private conversations (on the conversations similar to the on-domain corpus).

Our last experiment tests statistical association between health-related discussions in everyday telephone conversations, and actual medical events in older adults. As mentioned in Section 3, we have access to weekly self-reports on medical events for subjects' in everyday telephone conversations corpus. We used our best classifier, the SVM with union of expanded seed words, to classify all the conversations in our corpus into health-containing and health-free conversations. We then mark each conversation as temporally near a medical event if a reported medical event occurred within a 3-week time window. We chose a 3-week window to allow for one report before and after the event.

pain, **medicine**, appointment, **medical**, **doctors**, **emergency**, **prescription**, contact, **medication**, dial, **insurance**, **pharmacy**, schedule, moment, reached, questions, services, **surgery**, telephone, record, appointments, options, address, **patient**, advice, quality, tuesday, position, answered, records, wednesday, **therapy**, **healthy**, correct, department, ensure, numbers, act, **doctor**, personal, test, senior, **nurse**, plan, **kaiser**

Table 2: *50 most probable words in the topic of health returned by one run of TLDA. The bold words are the ones are hand-picked.*

Table 3 shows the number of conversations in different categories. At first glance it might seem like the number of false positives or false negatives is quite large but we should notice that being near a medical event is not the ground truth here. We just want to see if there is any association between occurrence of health-related conversations and occurrence of an actual medical event in lives of our subjects. We can see that 90.9%

of the conversations are classified as health-related but this percentage is slightly different for conversations near medical events(91.5%) vs. for the other conversations (89.1). This slight difference is significant according to χ^2 test of independence ($\chi^2(df = 1, N = 47288) = 61.17, p < 0.001$).

near a medical event	Classified as health-related	non-health-related
yes	1348	11067
no	2964	31909

Table 3: Number of telephone conversations in different categories. Each conversation is considered near a medical even if and only if there is at least one self-report in a window of 3 weeks around its date. Being near a medical event does not reveal the true nature of the conversation and thus is not the ground truth. So, there are no false positive, true positive, etc. in this table.

6 Conclusions

In this paper, we investigated the problem of identifying conversations with any mention of health. The private nature of our everyday telephone conversations corpus poses constraints on manual transcription and annotation. Looking up phone numbers associated with business calls, we labeled a small set of conversations when the other end was a business clearly related or unrelated to the health industry. However, the labeled set is not large enough for training a robust classifier. We developed a semi-supervised iterative method for selecting features, where we learn a distribution of words on health topic using TLDA, and subsequently filter irrelevant words iteratively. We demonstrate that our method generalizes well and improves the average F-score on in-domain and out-of-domain tasks over two baselines, using full vocabulary without feature selection or feature selection using TLDA alone. In our task, the generalization of the classifier is important since we are interested in detecting not only conversations on health with business (the annotated examples) but also with others in subjects' social network. Using our classifier, we find a significant statistical association between the occurrence of conversations about health and the occurrence of self-reported medical events.

Acknowledgments

This research was supported in part by NIH Grants 1K25AG033723, and P30 AG008017, as well as by NSF Grants 1027834, and 0964102. Any opinions, findings, conclusions or recommendations expressed in this publication are those of the authors and do not necessarily reflect the views of the NIH. We thank Nicole Larimer for help in collecting the data, Maider Lehr for testing the data collection devices and Katherine Wild for early discussions on this project. We are grateful to Brian Kingsbury and his colleagues for providing us access to IBM's *attila* software tools.

References

David Andrzejewski and Xiaojin Zhu. 2009. Latent dirichlet allocation with topic-in-set knowledge. In *Proceedings of the NAACL HLT 2009 Workshop on Semi-Supervised Learning for Natural Language Processing*, SemiSupLearn '09, pages 43–48.

David M. Blei, Andrew Y. Ng, and Michael I. Jordan. 2003. Latent dirichlet allocation. *Journal of Machine Learning Research*, 3:993–1022.

C.-C. Chang and C.-J. Lin. 2011. Libsvm : a library for support vector machines. *ACM Transactions on Intelligent Systems and Technology*, 2.

Christopher Cieri, David Miller, and Kevin Walker. 2004. The fisher corpus: a resource for the next generations of speech-to-text. In *LREC*, volume 4, pages 69–71.

John J Godfrey, Edward C Holliman, and Jane McDaniel. 1992. Switchboard: Telephone speech corpus for research and development. In *Acoustics, Speech, and Signal Processing, 1992. ICASSP-92., 1992 IEEE International Conference on*, volume 1, pages 517–520. IEEE.

Hong-qi Han, Dong-Hua Zhu, and Xue-feng Wang. 2009. Semi-supervised text classification from unlabeled documents using class associated words. In *Computers & Industrial Engineering, 2009. CIE 2009. International Conference on*, pages 1255–1260. IEEE.

Toru Hisamitsu and Yoshiki Niwa. 2001. Topic-word selection based on combinatorial probability. In *NLPRS*, volume 1, page 289.

Jagadeesh Jagarlamudi, Hal Daumé III, and Raghavendra Udupa. 2012. Incorporating lexical priors into topic models. In *Proceedings of the 13th Conference of the European Chapter of the Association for Computational Linguistics*, pages 204–213. Association for Computational Linguistics.

Soo-Min Kim and Eduard Hovy. 2004. Determining the sentiment of opinions. In *Proceedings of the 20th international conference on Computational Linguistics*, page 1367. Association for Computational Linguistics.

Hang Li and Kenji Yamanishi. 2003. Topic analysis using a finite mixture model. *Information processing & management*, 39(4):521–541.

Kyle W. Prier, Matthew S. Smith, Christophe Giraud-Carrier, and Carl L. Hanson. 2011. Identifying health-related topics on twitter: an exploration of tobacco-related tweets as a test topic. In *Proceedings of the 4th international conference on Social computing, behavioral-cultural modeling and prediction*, pages 18–25.

Hagen Soltau, Brian Kingsbury, Lidia Mangu, Daniel Povey, George Saon, and Geoffrey Zweig. 2005. The ibm 2004 conversational telephony system for rich transcription. In *Proc. ICASSP*, volume 1, pages 205–208.

Hong Yu and Vasileios Hatzivassiloglou. 2003. Towards answering opinion questions: Separating facts from opinions and identifying the polarity of opinion sentences. In *Proceedings of the 2003 conference on Empirical methods in natural language processing*, pages 129–136. Association for Computational Linguistics.

Taras Zagibalov and John Carroll. 2008. Automatic seed word selection for unsupervised sentiment classification of chinese text. In *Proceedings of the 22nd International Conference on Computational Linguistics-Volume 1*, pages 1073–1080. Association for Computational Linguistics.

Coreference Resolution for Structured Drug Product Labels

Halil Kilicoglu and Dina Demner-Fushman
National Library of Medicine
National Institutes of Health
Bethesda, MD, 20894
{kilicogluh,ddemner}@mail.nih.gov

Abstract

FDA drug package inserts provide comprehensive and authoritative information about drugs. DailyMed database is a repository of structured product labels extracted from these package inserts. Most salient information about drugs remains in free text portions of these labels. Extracting information from these portions can improve the safety and quality of drug prescription. In this paper, we present a study that focuses on resolution of coreferential information from drug labels contained in DailyMed. We generalized and expanded an existing rule-based coreference resolution module for this purpose. Enhancements include resolution of set/instance anaphora, recognition of appositive constructions and wider use of UMLS semantic knowledge. We obtained an improvement of 40% over the baseline with unweighted average F_1-measure using B-CUBED, MUC, and CEAF metrics. The results underscore the importance of set/instance anaphora and appositive constructions in this type of text and point out the shortcomings in coreference annotation in the dataset.

1 Introduction

Almost half of the US population uses at least one prescription drug and over 75% of physician office visits involve drug therapy[1]. Knowing how these drugs will affect the patient is very important, particularly, to over 20% of the patients that are on three or more prescription drugs[1]. FDA drug package inserts (drug labels or Structured Product Labels (SPLs)) provide curated information about the prescription drugs and many over-the-counter drugs. The drug labels for most drugs are publicly available in XML format through DailyMed [2]. Some information in these labels, such as the drug identifiers and ingredients, could be easily extracted from the structured fields of the XML documents. However, the salient content about indications, side effects and drug-drug interactions, among others, is buried in the free text of the corresponding sections of the labels. Extracting this information with natural language processing techniques can facilitate automatic timely updates to databases that support Electronic Health Records in alerting physicians to potential drug interactions, recommended doses, and contraindications.

Natural language processing methods are increasingly used to support various clinical and biomedical applications (Demner-Fushman et al., 2009). Extraction of drug information is playing a prominent role in these applications and research. In addition to earlier research in extraction of medications and relations involving medications from clinical text and the biomedical literature (Rindflesch et al., 2000; Cimino et al., 2007), in the third i2b2 shared task (Uzuner et al., 2010), 23 organizations have explored extraction of medications, their dosages, routes of administration, frequencies, durations, and reasons for administration from clinical text. The best performing systems used rule-based and machine learning techniques to achieve over 0.8 F-measure for extraction of medication names; however, the remaining information was harder to extract. Researchers have also tackled extraction of drug-drug interactions (Herrero-Zazo et al., 2013), side effects (Xu and Wang, 2014), and indications (Fung et al., 2013) from various biomedical resources.

As for many other information extraction tasks,

[1]Centers for Disease Control and Prevention: FASTSTATS - Therapeutic Drug Use: http://www.cdc.gov/nchs/fastats/drugs.htm

[2]DailyMed: http://dailymed.nlm.nih.gov/dailymed/about.cfm

Proceedings of the 2014 Workshop on Biomedical Natural Language Processing (BioNLP 2014), pages 45–53,
Baltimore, Maryland USA, June 26-27 2014.

extracting drug information is often made more difficult by coreference. Coreference is defined as the relation between linguistic expressions that are referring to the same entity (Zheng et al., 2011). Coreference resolution is a fundamental task in NLP and can benefit many downstream applications, such as relation extraction, summarization, and question answering. Difficulty of the task is due to the fact that various levels of linguistic information (lexical, syntactic, semantic, and discourse contextual features) generally play a role.

Coreference occurs frequently in all types of biomedical text, including the drug package inserts. Consider the example below:

(1) *Since amiodarone is a substrate for CYP3A and CYP2C8, drugs/substances that inhibit* ***these isoenzymes*** *may decrease the metabolism ....*

In this example, the expression *these isoenzymes* refer to *CYP3A* and *CYP2C8*. Resolving this coreference instance would allow us to capture the following drug interactions mentioned in the sentence: *inhibitors of CYP3A POTENTIATE amiodarone* and *inhibitors of CYP2C8 POTENTIATE amiodarone*.

In this paper, we present a study that focuses on identification of coreference links in drug labels, with the view that these relations will facilitate the downstream task of drug interaction recognition. The rule-based system presented is an extension of the previous work reported in Kilicoglu et al. (2013). The main focus of the dataset, based on SPLs, is drug interaction information. Coreference is only annotated when it is relevant to extracting such information. In addition to evaluating the system against a baseline, we also manually assessed the system output for precision. Furthermore, we also evaluated the system on a similarly drug-focused corpus annotated for anaphora (DrugNerAR) (Segura-Bedmar et al., 2010). Our results demonstrate that set/instance anaphora resolution and appositive recognition can play a significant role in this type of text and highlight some of the major areas of difficulty and potential enhancements.

2 Related Work

We discuss two areas of research related to this study in this section: processing of drug labels and coreference resolution focusing on biomedical text. Drug labels, despite their availability and the wealth of information contained within them, remain underutilized. One of the reasons might be the complexity of the text in the labels: in a review of publicly available text sources that could be used to augment a repository of drug indications and adverse effects (ADEs), Smith et al. (2011) concluded that many indication and adverse drug event relationships in the drug labels are too complex to be captured in the existing databases of interactions and ADEs. Despite the complexity, the labels were used to extract indications for drugs in several studies. Elkin et al. (2011) automatically extracted indications, mapped them to SNOMED-CT and then automatically derived rules in the form ("Drug" HasIndication "SNOMED CT"). Fung et al. (2013) used MetaMap (Aronson and Lang, 2010) to extract indications and map them to the UMLS (Lindberg et al., 1993), and then manually validated the quality of the mappings. Oprea et al. (2011) used information extracted from the adverse reactions sections of 988 drugs for computer-aided drug repurposing. Duke et al. (2011) have developed a rule-based system that extracted 534,125 ADEs from 5602 SPLs. Zhu et al. (2013) extracted disease terms from five SPL sections (indication, contraindication, ADE, precaution, and warning) and combined the extracted terms with the drug and disease relationships in NDF-RT to disambiguate the PharmGKB drug and disease associations. A hybrid NLP system, AutoMCExtractor, uses conditional random fields and post-processing rules to extract medical conditions from SPLs and build triplets in the form of([drug name]-[medical condition]-[LOINC section header]) (Li et al., 2013).

Coreference resolution in the biomedical domain was addressed in the 2011 i2b2/VA shared task (Uzuner et al., 2012), and the 2011 BioNLP Shared Task (Kim et al., 2012); however these community-wide evaluations did not change much the observation in the 2011 review by Zheng et al. (2011) that only a handful of systems were developed for handling anaphora and coreference in clinical text and biomedical publications. Since this comprehensive article was published, Yoshikawa et al. (2011) proposed two coference resolution models based on support vector machine and joint Markov logic network to aid the task of biological event extraction. Similarly, Miwa et al. (2012) and Kilicoglu and Bergler (2012) extended their biological event extraction pipelines using rule-based coreference systems that rely on syntactic information and predicate argument structures. Nguyen et al. (2012) evaluated contribution of discourse preference, number agreement, and domain-specific semantic information in capturing pronominal and nominal anaphora referring to proteins. An effort similar to ours is that of Segura-Bedmar et al. (2010), who resolve anaphora to support drug-drug interaction extraction. They created a corpus of 49 interactions sections extracted from the DrugBank database, having on average 40 sentences and 716 tokens. They then manually annotated pronominal and nominal anaphora, and developed a rule-based approach that achieve 0.76 F_1-measure in anaphora resolution.

3 Methods

3.1 The dataset

We used a dataset extracted from FDA drug package labels by our collaborators at FDA interested in extracting interactions between cardiovascular drugs. The dataset consists of 159 drug labels, with an average of 105 sentences and 1787 tokens per label. It is annotated for three entity types (Drug, Drug Class, and Substance) and four drug interaction types (Caution, Decrease, Increase, and Specific). 377 instances of coreference were annotated. Two annotators separately annotated the labels and one of the authors performed the adjudication. The relatively low number of coreference instances is due to the fact that coreference was annotated only when it would be relevant to drug interaction recognition task. This parsimonious approach to annotation presents difficulty in automatically evaluating the system, and to mitigate this, we present an assessment of the precision of our end-to-end coreference system, as well. We split the dataset into training and test sets by random sampling. Training data consists of 79 documents and the test set has 80 documents. We used the training data for analysis and as the basis of our enhancements.

3.2 The system

The work described in this paper extends and refines earlier work, described in Kilicoglu et al. (2013), which focused on disease anaphora and ellipsis in the context of consumer health questions. We briefly recap that system here. The system begins by mapping named entities to UMLS Metathesaurus concepts (CUIs). Next, it identifies anaphoric expressions in text, which include personal (e.g., *it*, *they*) and demonstrative pronouns (e.g., *this*, *those*), as well as sortal anaphora (definite (e.g., with *the*) and demonstrative (e.g., with *that*) noun phrases). The candidate antecedents are then recognized using syntactic (person, gender and number agreement, head word matching) and semantic (hypernym and UMLS semantic type matching) constraints. Finally, the co-referent is then selected as the *focus* of the question, which is taken as the first disease mention in the question.

The coreference resolution pipeline used in the current work, while enhanced significantly, follows the same basic sequence. The relatively simple approach of earlier work is generally sufficient for consumer health questions; however, we found it insufficient when it comes to drug labels. Aside from the obvious point that the approach was limited to diseases, there are other stylistic differences that have an impact on coreference resolution. In contrast to informal and casual style of consumer health questions, drug labels are curated and provide complex indication and ADE information in a formal style, more akin to biomedical literature. Our analysis of the training data highlighted several facts regarding coreference in drug labels: (1) the set/instance anaphora (including those involving distributive anaphora such as *both*, *each*, *either*) instances are prominent, (2) demonstrative pronominal anaphora is non-existent in contrast to consumer health questions, (3) the *focus*-based salience scoring is simplistic for longer texts. We describe the system enhancements below.

3.2.1 Generalizing from diseases to drugs and beyond

We generalized from resolution of disease coreference only to resolution of coreference involving other entity types. For this purpose, we parameterized semantic groups and hypernym lists associated with each semantic group. We generalized the system in the sense that new semantic types and hypernyms can be easily defined and used by the system. In addition to Disorder semantic group and Disorder hypernym list defined in earlier work, we used Drug, Intervention, Population, Procedure, Anatomy, and Gene/Protein semantic groups and hypernym lists. Semantic group classification largely mimics coarse-grained UMLS semantic groups (McCray et al., 2001). For example, UMLS semantic types Pharmacologic Substance and Clinical Drug are aggregated into both Drug and Intervention semantic groups, while Therapeutic or Preventive Procedure is assigned to Procedure group only. Drug hypernyms, such as *medication*, *drug*, *agent*, were derived from the training data.

3.2.2 Set/instance anaphora

Set/instance anaphora instances are prevalent in drug labels. In our dataset, 19% of all annotated anaphoric expressions indicate set/instance anaphora (co-referring with 29% of antecedent terms). An example was provided earlier (Example 1). While recognizing anaphoric expressions that indicate set/instance anaphora is not necessarily difficult (i.e., recognizing *these isoenzymes* in the example), linking them to their antecedents can be difficult, since it generally involves correctly identifying syntactic coordination, a challenging syntactic parsing task (Ogren, 2010). Our identification of these structures relies on collapsed Stanford dependency output (de Marneffe et al., 2006) and uses syntactic and semantic constraints. We examine all the dependency relations extracted from a sentence and only consider those with the type *conj_** (e.g., *conj_and*, *conj_or*). For increased accuracy, we then check the tokens involved in the dependency (conjuncts) and ensure that there is a coordinating conjunction (e.g., *and*, *or*, , (comma), & (ampersand)) between them. Once such a conjunction is identified, we then examine the semantic compatibility of the conjuncts. In the case of entities, the compatibility involves that at the semantic group level. In the current work, we also began recognizing distributive anaphora, such as *either*, *each* as anaphoric expressions. When the recognized anaphoric expression is plural (as in *they*, *these agents* or *either drug*), we allow the coordinated structures previously identified in this fashion as candidate antecedents. The current work does not address a more complex kind of set/instance anaphora, in which the instances are not syntactically coordinated, such as in Example (2), where *such agents* refer to *thiazide diuretics*, in the preceding sentence, as well as *Potassium-sparing diuretics* and *potassium supplements*.

(2) *...can attenuate potassium loss caused by thiazide diuretics. Potassium-sparing diuretics ...or potassium supplements can increase .... if concomitant use of* ***such agents*** *is indicated...*

3.2.3 Appositive constructions

Coreference involving appositive constructions[3] are annotated in some corpora, including the BioNLP shared task coreference dataset (Kim et al., 2012) and DrugNerAR corpus (Segura-Bedmar et al., 2010). An example is given below, in which the indefinite noun phrase *a drug* and the drug *lovastatin* are appositives.

(3) *PLETAL does not, however, appear to cause increased blood levels of drugs metabolized by CYP3A4, as it had no effect on lovastatin,* ***a drug*** *with metabolism very sensitive to CYP3A4 inhibition.*

In our dataset, coreference involving appositive constructions were generally left unannotated. However, it was consistently the case that when one of the items in the construction is annotated as the antecedent for an anaphoric expression, the other item in the construction was also annotated as such. Therefore, we identified appositive constructions in text to aid the antecedent selection task. We used dependency relations for this task, as well. Identifying appositives is relatively straightforward using syntactic dependency relations. We adapted the following rule from Kilicoglu and Bergler (2012):

$$\mathit{APPOS(Antecedent,Anaphor)} \vee \mathit{APPOS(Anaphor,Antecedent)} \Rightarrow \mathit{COREF(Anaphor,Antecedent)}$$

where $\mathit{APPOS} \in \{\mathit{appos}, \mathit{abbrev}, \mathit{prep\_including}, \mathit{prep\_such\_as}\}$. In our case, this rule becomes

$$(\mathit{APPOS(Antecedent1,Antecedent2)} \vee \mathit{APPOS(Antecedent2,Antecedent1)}) \wedge \mathit{COREF(Anaphor,Antecedent1)} \Rightarrow \mathit{COREF(Anaphor,Antecedent2)}$$

which essentially states that a candidate is taken as an antecedent, only if its appositive has been recognized as an antecedent. Additionally, semantic compatibility between the items is required.

This allows us to identify *their* and *Class Ia antiarrhythmic drugs* as co-referents in the following example, due to the fact that the exemplification indicated by the appositive construction between *Class Ia antiarrythmic drugs* and *disopyramide* is recognized, the latter previously identified as an antecedent for *their*.

[3]We use the term "appositive" to cover exemplifications, as well.

(4) *Class Ia antiarrhythmic drugs, such as disopyramide, quinidine and procainamide and other Class III drugs (e.g., amiodarone) are not recommended ... because of* ***their*** *potential to prolong refractoriness.*

3.2.4 Relative pronouns

Similar to appositive constructions, relative pronouns are annotated as anaphoric expressions in some corpora (same as those for appositives), but not in our dataset. In the example below, the relative pronoun *which* refers to *potassium-containing salt substitutes*.

(5) *... the concomitant use of potassium-sparing diuretics, potassium supplements, and/or potassium-containing salt substitutes,* ***which*** *should be used cautiously...*

Since we aim for generality and this type of anaphora can be important for downstream applications, we implemented a rule, again taken from Kilicoglu and Bergler (2012), which simply states that the antecedent of a relative pronominal anaphora is the noun phrase head it modifies.

$$rel(X,Anaphor) \wedge rcmod(Antecedent,X) \Rightarrow COREF(Anaphor,Antecedent)$$

where *rel* indicates a *relative* dependency, and *rcmod* a *relative clause modifier* dependency. We extended this in the current work to include the following rules:

(6) (a) $LEFT(Antecedent,Anaphor) \wedge NO\_INT\_WORD(Antecedent,Anaphor) \Rightarrow COREF(Anaphor,Antecedent)$

(b) $LEFT(Antecedent,Anaphor) \wedge rcmod(Antecedent,X) \wedge LEFT(Anaphor,X) \Rightarrow COREF(Anaphor,Antecedent)$

where *LEFT* indicates that the first argument is to the left of the second and *NO_INT_WORD* indicates that the arguments have no intervening words between them.

3.3 Drug ingredient/brand name synonymy

A specific, non-anaphoric type of coreference, between drug ingredient name and drug's brand name, is commonly annotated in our dataset. An example is provided below, where *COREG CR* is the brand name for *carvedilol*.

(7) *The concomitant administration of amiodarone or other CYP2C9 inhibitors such as fluconazole with COREG CR may enhance the -blocking properties of* ***carvedilol*** *....*

To identify this type of coreference, we use semantic information from UMLS Metathesaurus. We stipulate that, to qualify as co-referents, both terms under consideration should map to the same UMLS concept (i.e., that they are considered synonyms). If the terms are within the same sentence, we further require that they are appositive.

3.3.1 Demonstrative pronouns

Anaphoric expressions of demonstrative pronoun type generally have discourse-deictic use; in other words, they often refer to events, propositions described in prior discourse or even to the full sentences or paragraphs, rather than concrete objects or entities (Webber, 1988). This fact was implicitly exploited in consumer health questions, since the coreference resolution focused on diseases only, which are essentially processes. However, in drug labels, discourse-deictic use of demonstratives is much more overt. Consider the sentence below, where the demonstrative *This* refers to the event of *increasing the exposure to lovastatin*.

(8) *Co-administration of lovastatin and SAMSCA increases the exposure to lovastatin and ....* ***This*** *is not a clinically relevant change.*

To handle such cases, we blocked entity antecedents (such as drugs) for demonstrative pronouns and only allowed predicates (verbs, nominalizations) as candidate antecedents.

3.3.2 Pleonastic *it*

We recognized pleonastic instances of the pronoun *it* to disqualify them as anaphoric expressions (for instance, *it* in *It may be necessary to ...*). Generally, lexical patterns involving sequence of tokens are used to recognize such instances (e.g., (Segura-Bedmar et al., 2010). We used a simple dependency-based rule that mimics these patterns, given below.

$$nsubj^{*}(X,it) \wedge DEP(X,Y) \Rightarrow PLEONASTIC(it)$$

where *nsubj** refers to *nsubj* or *nsubjpass* dependencies and *DEP* is any dependency, where $DEP \notin \{infmod, ccomp, xcomp\}$.

3.3.3 Discourse-based constraints

Previously, we did not impose limits on how far the co-referents could be from each other, since the entire discourse was generally short and the salient antecedent (often the topic of the question) appeared early in discourse. This is often not the case in drug labels, especially because often intricate interactions between the drug of interest and other medications are discussed. Therefore, we limit the discourse window from which candidate antecedents are identified. Generally, the search space for the antecedents is limited to the current sentence as well as the two preceding sentences (Segura-Bedmar et al., 2010; Nguyen et al., 2012). In our dataset, we found that 98% of antecedents occurred within this discourse window and, thus, use the same search space. We make an exception for the cases in which the anaphoric expression appear in the first sentence of a paragraph and no compatible antecedent is found in the same sentence. In this case, the search space is expanded to the entire preceding paragraph.

We also extended the system to include different types of salience scoring methods. For drug labels, we use linear distance between the co-referents (in terms of surface elements) as the salience score; the lower this score, the better candidate the antecedent is. Additionally, we implemented syntactic tree distance between the co-referents as a potential salience measure, even though this type of salience scoring did not have an effect on our results on drug labels.

Finally, we block candidate antecedents that are in a direct syntactic dependency with the anaphoric expression, except when the anaphor is reflexive (e.g., *itself*).

3.4 Evaluation

To evaluate our approach, we used a baseline similar to that reported in Segura-Bedmar et al. (2010), which consists of selecting the closest preceding nominal phrase for the anaphoric expressions annotated in their corpus. These expressions include pronominal (personal, relative, demonstrative, etc.) and nominal (definite, possessive, etc.) anaphora. We compared our system to this baseline using the unweighted average of F_1-measure over B-CUBED (Bagga and Baldwin, 1998), MUC (Vilain et al., 1995), and CEAF (Luo, 2005) metrics, the standard evaluation metrics for coreference resolution. We used the scripts provided by i2b2 shared task organizers for this purpose. Since coreference annotation was parsimonious in our dataset, we also manually examined a subset of the coreference relations extracted by the system for precision. Additionally, we tested our system on DrugNerAR corpus (Segura-Bedmar et al., 2010), which similarly focuses on drug interactions. We compared our results to theirs, using as evaluation metrics precision, recall, and F_1-measure, the metrics that were used in their evaluation.

4 Results and Discussion

With the drug label dataset, we obtained the best results without relative pronominal anaphora resolution and drug ingredient/brand name synonymy strategies (OPTIMAL) and with linear distance as the salience measure. In this setting, using gold entity annotations, we recognized 318 coreference chains, 54 of which were annotated in the corpus. The baseline identified 1415 coreference chains, only 10 of which were annotated. The improvement provided by the system over the baseline is clear; however, the low precision/recall/F_1-measure, given in Table 1, should be taken with caution due to the sparse coreference annotation in the dataset. To get a better sense of how well our system performs, we also performed end-to-end coreference resolution and manually assessed a subset of the system output (22 randomly selected drug labels with 249 coreference instances). Of these 249, 181 were deemed correct, yielding a precision of 0.73. The baseline method extracted 1439 instances, 56 of which were deemed correct, yielding a precision of 0.04. The precision of our method is more in line with what has been reported in the literature (Segura-Bedmar et al., 2010; Nguyen et al., 2012). For i2b2-style evaluation using the unweighted average F_1 measure over B-CUBED, MUC, and CEAF metrics, we considered both exact and partial mention overlap. These results, provided in Table 1, also indicate that the system provides a clear improvement over the baseline.

Metric	Baseline	OPTIMAL
With gold entity annotations		
Unweighted F_1 Partial	0.55	0.77
Unweighted F_1 Exact	0.66	0.78
Precision	0.01	0.17
Recall	0.04	0.26
F_1-measure	0.01	0.21
End-to-end coreference resolution		
Precision	0.04	0.73

Table 1: Evaluation results on drug labels

We also assessed the effect of various resolution strategies on results. These results are presented in Table 2.

Strategy	**F_1-measure**
OPTIMAL	0.21
OPTIMAL - SIA	0.21
OPTIMAL - APPOS	0.15
OPTIMAL + DIBS	0.16 (0.39 recall)

Table 2: Effect of coreference strategies

Disregarding set/instance anaphora resolution (SIA) does not appear to affect the results by much; however, this is mostly due to the fact that the "instance" mentions are generally exemplifications of a particular drug class which also appear in text. In the absence of set/instance anaphora resolution, the system often defaults to these drug class mentions, which were annotated more often than not, unlike the "instance" mentions. Take the following example:

(9) *Use of ZESTRIL with potassium-sparing diuretics (e.g., spironolactone, eplerenone, triamterene or amiloride)... may lead to significant increases ... if concomitant use of **these agents** ...*

Without set-instance anaphora resolution, the system links *these agents* to *potassium-sparing diuretics*, an annotated relation. With set-instance anaphora resolution, the same expression is linked to individual drug names (*spironolactone*, etc.) as well as the the drug class, creating a number of false positives, which, in effect, offsets the improvement provided by this strategy.

On the other hand, recognizing appositive constructions (APPOS) appears to have a larger impact; however, it should be noted that this is mostly because it helps us expand the antecedent mention list in the case of set/instance anaphora. For instance, in Example (9), this strategy allows us to establish the link between the anaphora and the drug class (*diuretics*), since the drug class and individual drug name (*spironolactone*) are identified earlier as appositive. We can conclude that, in general, set/instance anaphora benefits from recognition of appositive constructions.

Recognizing drug ingredient/brand name synonymy (DIBS) improved the recall and hurt the precision significantly, the overall effect being negative. Since this non-anaphoric type of coreference is strictly semantic in nature and resources from which this type of semantic information can be derived already exist (UMLS, among others), it is perhaps not of utmost importance that a coreference resolution system recognizes such coreference.

We additionally processed the DrugNerAR corpus with our system. The optimal setting for this corpus was disregarding the drug ingredient/brand name synonymy but using relative pronoun anaphora resolution, based on the discussion in Segura-Bedmar et al. (2010). Somewhat to our surprise, our system did not fare well on this corpus. We extracted 524 chains, 327 of which (out of 669) were annotated in the corpus, yielding a precision of 0.71, recall of 0.56, and F_1-measure of 0.63. This is about 20% lower than their reported results. When we used their baseline method (explained earlier), we obtained similarly lower scores (precision of 0.18, recall of 0.45, F_1-measure of 0.26, about 40% lower than their reported results). In light of this apparent discrepancy, which clearly warrants further investigation, it is perhaps more sensible to focus on "improvement over baseline" (reported as 73% in their paper and is 140% in our case).

We analyzed some of the annotations more closely to get a better sense of the shortcomings of the system. The majority of errors were due to using linear distance as the salience score. For instance, in the following example, *they* is linked to *ACE inhibitors* due to proximity, whereas the true antecedent is *these reactions* (itself an anaphor and is presumably linked to another antecedent). It could be possible to recover this link using principles of Centering Theory (Grosz et al., 1995), which suggests that subjects are more central than objects and adjuncts in an utterance. Following this principle, the subject (*these reactions*) would be preferred to *ACE inhibitors* as the antecedent.

(10) *In the same patients, these reactions were avoided when ACE inhibitors were temporarily withheld, but **they** reappeared upon inadvertent rechallenge.*

Semantic (but not syntactic) coordination sometimes leads to number disagreement between the anaphora and a true antecedent, as shown in Example (11), leading to false negatives. In this example, *such diuretics* refers to both *ALDACTONE* and *a second diuretic*; however, we are unable to identify the link between them and the number disagreement between the anaphora and either of the antecedents blocks a potential coreference relation between these items.

(11) *If, after five days, an adequate diuretic response to ALDACTONE has not occurred, a second diuretic that acts more proximally in the renal tubule may be added to the regimen. Because of the additive effect of ALDACTONE when administered concurrently with* ***such diuretics*** ...

5 Conclusion

We presented a coreference resolution system enhanced based on insights from a dataset of FDA drug package inserts. Sparse coreference annotation in the dataset presented difficulties in evaluating the results; however, based on various evaluation strategies, the performance improvement due to the enhancements seems evident. Our results show that recognizing coordination and appositive constructions are particularly useful and that non-anaphoric cases of coreference can be identified using synonymy in semantic resources, such as UMLS. However, whether this is a task for a coreference resolution system or a concept normalization system is debatable. We experimented with using hierarchical domain knowledge in UMLS (for example, the knowledge that *lisinopril* ISA *angiotensin converting enzyme inhibitor*) to resolve some cases of sortal anaphora. Even though we did not see an improvement due to using this type of information on our dataset, further work is needed to assess its usefulness. While the enhancements were evaluated on drug labels only, they are not specific to this type of text. Their portability to different text types is limited only by the accuracy of underlying tools, such as parsers, for the text type of interest and the availability of domain knowledge in the form of relevant semantic types, groups, hypernyms for the entity types under consideration. The results also indicate that a more rigorous application of syntactic constraints in the spirit of Centering Theory (Grosz et al., 1995) could be beneficial. Event (or clausal) anaphora and anaphora indicating discourse deixis, while rarely annotated in our dataset, appear to occur fairly often in biomedical text. These types of anaphora are known to be particularly challenging, and we plan to investigate them in future research, as well.

Acknowledgments

This work was supported by the intramural research program at the U.S. National Library of Medicine, National Institutes of Health.

References

Alan R. Aronson and François-Michel Lang. 2010. An overview of MetaMap: historical perspective and recent advances. *Journal of the American Medical Informatics Association (JAMIA)*, 17(3):229–236.

Amit Bagga and Breck Baldwin. 1998. Algorithms for scoring coreference chains. In *The First International Conference on Language Resources and Evaluation Workshop on Linguistics Coreference*, pages 563–566.

James J. Cimino, Tiffani J. Bright, and Jianhua Li. 2007. Medication reconciliation using natural language processing and controlled terminologies. In Klaus A. Kuhn, James R. Warren, and Tze-Yun Leong, editors, *MedInfo*, volume 129 of *Studies in Health Technology and Informatics*, pages 679–683. IOS Press.

Marie-Catherine de Marneffe, Bill MacCartney, and Christopher D. Manning. 2006. Generating typed dependency parses from phrase structure parses. In *Proceedings of the 5th International Conference on Language Resources and Evaluation*, pages 449–454.

Dina Demner-Fushman, Wendy W. Chapman, and Clem J. McDonald. 2009. What can natural language processing do for clinical decision support? *Journal of Biomedical Informatics*, 5(42):760–762.

Jon Duke, Jeff Friedlin, and Patrick Ryan. 2011. A quantitative analysis of adverse events and "overwarning" in drug labeling. *Archives of internal medicine*, 10(171):944–946.

Peter L. Elkin, John S. Carter, Manasi Nabar, Mark Tuttle, Michael Lincoln, and Steven H. Brown. 2011. Drug knowledge expressed as computable semantic triples. *Studies in health technology and informatics*, (166):38–47.

Kin Wah Fung, Chiang S. Jao, and Dina Demner-Fushman. 2013. Extracting drug indication information from structured product labels using natural language processing. *JAMIA*, 20(3):482–488.

Barbara J. Grosz, Scott Weinstein, and Aravind K. Joshi. 1995. Centering: a framework for modeling the local coherence of discourse. *Computational Linguistics*, 21(2):203–225.

María Herrero-Zazo, Isabel Segura-Bedmar, Paloma Martínez, and Thierry Declerck. 2013. The DDI corpus: An annotated corpus with pharmacological substances and drug-drug interactions. *Journal of Biomedical Informatics*, 46(5):914–920.

Halil Kilicoglu and Sabine Bergler. 2012. Biological Event Composition. *BMC Bioinformatics*, 13 (Suppl 11):S7.

Halil Kilicoglu, Marcelo Fiszman, and Dina Demner-Fushman. 2013. Interpreting consumer health questions: The role of anaphora and ellipsis. In *Proceedings of the 2013 Workshop on Biomedical Natural Language Processing*, pages 54–62.

Jin-Dong Kim, Ngan Nguyen, Yue Wang, Jun'ichi Tsujii, Toshihisa Takagi, and Akinori Yonezawa. 2012. The Genia Event and Protein Coreference tasks of the BioNLP Shared Task 2011. *BMC Bioinformatics*, 13(Suppl 11):S1.

Qi Li, Louise Deleger, Todd Lingren, Haijun Zhai, Megan Kaiser, Laura Stoutenborough, Anil G. Jegga, Kevin B. Cohen, and Imre Solti. 2013. Mining FDA drug labels for medical conditions. *BMC medical informatics and decision making*, 13(1):53.

Donald A. B. Lindberg, Betsy L. Humphreys, and Alexa T. McCray. 1993. The Unified Medical Language System. *Methods of Information in Medicine*, 32:281–291.

Xiaoqiang Luo. 2005. On coreference resolution performance metrics. In *In Proc. of HLT/EMNLP*, pages 25–32.

Alexa T. McCray, Anita Burgun, and Olivier Bodenreider. 2001. Aggregating UMLS semantic types for reducing conceptual complexity. *Proceedings of Medinfo*, 10(pt 1):216–20.

Makoto Miwa, Paul Thompson, and Sophia Ananiadou. 2012. Boosting automatic event extraction from the literature using domain adaptation and coreference resolution. *Bioinformatics*, 28(13):1759–1765.

Ngan L. T. Nguyen, Jin-Dong Kim, Makoto Miwa, Takuya Matsuzaki, and Junichi Tsujii. 2012. Improving protein coreference resolution by simple semantic classification. *BMC Bioinformatics*, 13:304.

Philip V. Ogren. 2010. Improving Syntactic Coordination Resolution using Language Modeling. In *NAACL (Student Research Workshop)*, pages 1–6. The Association for Computational Linguistics.

T.I. Oprea, S.K. Nielsen, O. Ursu, J.J. Yang, O. Taboureau, S.L. Mathias, L. Kouskoumvekaki, L.A. Sklar, and C.G. Bologa. 2011. Associating Drugs, Targets and Clinical Outcomes into an Integrated Network Affords a New Platform for Computer-Aided Drug Repurposing. *Molecular informatics*, 2-3(30):100–111.

Thomas C. Rindflesch, Lorrie Tanabe, John N. Weinstein, and Lawrence Hunter. 2000. EDGAR: Extraction of drugs, genes, and relations from the biomedical literature. In *Proceedings of Pacific Symposium on Biocomputing*, pages 514–525.

Isabel Segura-Bedmar, Mario Crespo, César de Pablo-Sánchez, and Paloma Martínez. 2010. Resolving anaphoras for the extraction of drug-drug interactions in pharmacological documents. *BMC Bioinformatics*, 11 (Suppl 2):S1.

J.C. Smith, J.C. Denny, Q. Chen, H. Nian, A. 3rd Spickard, S.T. Rosenbloom, and R. A. Miller. 2011. Lessons learned from developing a drug evidence base to support pharmacovigilance. *Applied clinical informatics*, 4(4):596–617.

Özlem Uzuner, Imre Solti, and Eithon Cadag. 2010. Extracting medication information from clinical text. *JAMIA*, 17(5):514–518.

Özlem Uzuner, Andrea Bodnari, Shuying Shen, Tyler Forbush, John Pestian, and Brett R. South. 2012. Evaluating the state of the art in coreference resolution for electronic medical records. *JAMIA*, 19(5):786–791.

Marc B. Vilain, John D. Burger, John S. Aberdeen, Dennis Connolly, and Lynette Hirschman. 1995. A model-theoretic coreference scoring scheme. In *MUC*, pages 45–52.

Bonnie L. Webber. 1988. Discourse Deixis: Reference to Discourse Segments. In *ACL*, pages 113–122.

Rong Xu and QuanQiu Wang. 2014. Large-scale combining signals from both biomedical literature and the FDA Adverse Event Reporting System (FAERS) to improve post-marketing drug safety signal detection. *BMC Bioinformatics*, 15:17.

Katsumasa Yoshikawa, Sebastian Riedel, Tsutomu Hirao, Masayuki Asahara, and Yuji Matsumoto. 2011. Coreference Based Event-Argument Relation Extraction on Biomedical Text. *Journal of Biomedical Semantics*, 2 (Suppl 5):S6.

Jiaping Zheng, Wendy W. Chapman, Rebecca S. Crowley, and Guergana K. Savova. 2011. Coreference resolution: A review of general methodologies and applications in the clinical domain. *Journal of Biomedical Informatics*, 44(6):1113–1122.

Qian Zhu, Robert R. Freimuth, Jyotishman Pathak, Matthew J. Durski, and Christopher G. Chute. 2013. Disambiguation of PharmGKB drug-disease relations with NDF-RT and SPL. *Journal of Biomedical Informatics*, 46(4):690–696.

Generating Patient Problem Lists from the ShARe Corpus using SNOMED CT/SNOMED CT CORE Problem List

Danielle Mowery
Janyce Wiebe
University of Pittsburgh
Pittsburgh, PA
dlm31@pitt.edu
wiebe@cs.pitt.edu

Mindy Ross
University of California
San Diego
La Jolla, CA
mkross@ucsd.edu

Sumithra Velupillai
Stockholm University
Stockholm, SE
sumithra@dsv.su.se

Stephane Meystre
Wendy W Chapman
University of Utah
Salt Lake City, UT
stephane.meystre,
wendy.chapman@utah.edu

Abstract

An up-to-date problem list is useful for assessing a patient's current clinical status. Natural language processing can help maintain an accurate problem list. For instance, a patient problem list from a clinical document can be derived from individual problem mentions within the clinical document once these mentions are mapped to a standard vocabulary. In order to develop and evaluate accurate document-level inference engines for this task, a patient problem list could be generated using a standard vocabulary. Adequate coverage by standard vocabularies is important for supporting a clear representation of the patient problem concepts described in the texts and for interoperability between clinical systems within and outside the care facilities. In this pilot study, we report the reliability of domain expert generation of a patient problem list from a variety of clinical texts and evaluate the coverage of annotated patient problems against SNOMED CT and SNOMED **C**linical **O**bservation **R**ecording and **E**ncoding (CORE) Problem List. Across report types, we learned that patient problems can be annotated with agreement ranging from 77.1% to 89.6% F1-score and mapped to the CORE with moderate coverage ranging from 45%-67% of patient problems.

1 Introduction

In the late 1960's, Lawrence Weed published about the importance of problem-oriented medical records and the utilization of a problem list to facilitate care provider's clinical reasoning by reducing the cognitive burden of tracking *current, active* problems from *past, inactive* problems from the patient health record (Weed, 1970). Although electronic health records (EHR) can help achieve better documentation of problem-specific information, in most cases, the problem list is manually created and updated by care providers. Thus, the problem list can be out-of-date containing resolved problems or missing new problems. Providing care providers with problem list update suggestions generated from clinical documents can improve the completeness and timeliness of the problem list (Meystre and Haug, 2008).

In recent years, national incentive and standard programs have endorsed the use of problem lists in the EHR for tracking patient diagnoses over time. For example, as part of the Electronic Health Record Incentive Program, the Center for Medicare and Medicaid Services defined demonstration of *Meaningful Use* of adopted health information technology in the Core Measure 3 objective as "maintaining an up-to-date problem list of current and active diagnoses in addition to historical diagnoses relevant to the patients care" (Center for Medicare and Medicaid Services, 2013). More recently, the **S**ystematized **N**omenclature of **M**edicine **C**linical **T**erms (SNOMED CT) has become the standard vocabulary for representing and documenting patient problems within the clinical record. Since 2008, this list is iteratively refined four times each year to produce a subset of generalizable clinical problems called the SNOMED CT CORE Problem List. This CORE list represents the most frequent problem terms and concepts across eight major healthcare institutions in the United States and is designed to support interoperability between regional healthcare institutions (National Library of Medicine, 2009).

In practice, there are several methodologies applied to generate a patient problem list from clinical text. Problem lists can be generated from coded diagnoses such as the **I**nternational **S**tatistical **C**lassification of **D**isease (ICD-9 codes) or

Proceedings of the 2014 Workshop on Biomedical Natural Language Processing (BioNLP 2014), pages 54–58, Baltimore, Maryland USA, June 26-27 2014. ©2014 Association for Computational Linguistics

concept labels such as **U**nified **M**edical **L**anguage **S**ystem **c**oncept **u**nique **i**dentifiers (UMLS CUIs). For example, Meystre and Haug (2005) defined 80 of the most frequent problem concepts from coded diagnoses for cardiac patients. This list was generated by a physician and later validated by two physicians independently. Coverage of coded patient problems were evaluated against the ICD-9-CM vocabulary. Solti et al. (2008) extended the work of Meystre and Haug (2005) by not limiting the types of patient problems from any list or vocabulary to generate the patient problem list. They observed 154 unique problem concepts in their reference standard. Although both studies demonstrate valid methods for developing a patient problem list reference standard, neither study leverages a standard vocabulary designed specifically for generating problem lists.

The goals of this study are 1) determine how reliably two domain experts can generate a patient problem list leveraging SNOMED CT from a variety of clinical texts and 2) assess the coverage of annotated patient problems from this corpus against the CORE Problem List.

2 Methods

In this IRB-approved study, we obtained the **Sh**ared **A**nnotated **Re**source (ShARe) corpus originally generated from the Beth Israel Deaconess Medical Center (Elhadad et al., under review) and stored in the **M**ultiparameter **I**ntelligent **M**onitoring in **I**ntensive **C**are, version 2.5 (MIMIC II) database (Saeed et al., 2002). This corpus consists of discharge summaries (DS), radiology (RAD), electrocardiogram (ECG), and echocardiogram (ECHO) reports from the **I**ntensive **C**are **U**nit (ICU). The ShARe corpus was selected because it 1) contains a variety of clinical text sources, 2) links to additional patient structured data that can be leveraged for further system development and evaluation, and 3) has encoded individual problem mentions with semantic annotations within each clinical document that can be leveraged to develop and test document-level inference engines. We elected to study ICU patients because they represent a sensitive cohort that requires up-to-date summaries of their clinical status for providing timely and effective care.

2.1 Annotation Study

For this annotation study, two annotators - a physician and nurse - were provided independent training to annotate clinically relevant problems e.g., *signs*, *symptoms*, *diseases*, and *disorders*, at the document-level for 20 reports. The annotators were given feedback based on errors over two iterations. For each patient problem in the remaining set, the physician was instructed to review the full text, span the a problem mention, and map the problem to a CUI from SNOMED-CT using the extensible Human Oracle Suite of Tools (eHOST) annotation tool (South et al., 2012). If a CUI did not exist in the vocabulary for the problem, the physician was instructed to assign a "CUI-less" label. Finally, the physician then assigned one of five possible status labels - *Active*, *Inactive*, *Resolved*, *Proposed*, and *Other* - based on our previous study (Mowery et al., 2013) to the mention representing its last status change at the conclusion of the care encounter. Patient problems were not annotated as *Negated* since patient problem concepts are assumed absent at a document-level (Meystre and Haug, 2005). If the patient was healthy, the physician assigned "Healthy - no problems" to the text. To reduce the cognitive burden of annotation and create a more robust reference standard, these annotations were then provided to a nurse for review. The nurse was instructed to add missing, modify existing, or delete spurious patient problems based on the guidelines.

We assessed how reliably annotators agreed with each other's patient problem lists using inter-annotator agreement (IAA) at the document-level. We evaluated IAA in two ways: 1) by problem CUI and 2) by problem CUI and status. Since the number of problems not annotated (i.e., ***true negatives (TN)***) are very large, we calculated F1-score as a surrogate for kappa (Hripcsak and Rothschild, 2005). F1-score is the harmonic mean of recall and precision, calculated from ***true positive***, ***false positive***, and ***false negative*** annotations, which were defined as follows:

true positive (TP) = the physician and nurse problem annotation was assigned the same CUI (and status)

false positive (FP) = the physician problem annotation (and status) did not exist among the nurse problem annotations

false negative (FN) = the nurse problem annotation (and status) did not exist among the physician problem annotations

Recall =

$$\frac{TP}{(TP + FN)} \quad (1)$$

Precision =

$$\frac{TP}{(TP + FP)} \quad (2)$$

F1-score =

$$2\frac{(Recall * Precision)}{(Recall + Precision)} \quad (3)$$

We sampled 50% of the corpus and determined the most common errors. These errors with *examples* were programmatically adjudicated with the following **solutions**:

Spurious problems: *procedures*
solution: exclude non-problems via guidelines

Problem specificity: *CUI specificity differences*
solution: select most general CUIs

Conflicting status: *negated vs. resolved*
solution: select second reviewer's status

CUI/CUI-less: *C0031039 vs. CUI-less*
solution: select CUI since clinically useful

We split the dataset into about two-thirds training and one-third test for each report type. The remaining data analysis was performed on the training set.

2.2 Coverage Study

We characterized the composition of the reference standard patient problem lists against two standard vocabularies SNOMED-CT and SNOMED-CT CORE Problem List. We evaluated the coverage of patient problems against the SNOMED CT CORE Problem List since the list was developed to support encoding clinical observations such as findings, diseases, and disorders for generating patient summaries like problem lists. We evaluated the coverage of patient problems from the corpus against the SNOMED-CT January 2012 Release which leverages the UMLS version 2011AB. We assessed recall (Eq 1), defining a TP as a patient problem CUI occurring in the vocabulary and a FN as a patient problem CUI not occurring in the vocabulary.

3 Results

We report the results of our annotation study on the full set and vocabulary coverage study on the training set.

3.1 Annotation Study

The full dataset is comprised of 298 clinical documents - 136 (45.6%) DS, 54 (18.1%) ECHO, 54 (18.1%) RAD, and 54 (18.1%) ECG. Seventy-four percent (221) of the corpus was annotated by both annotators. Table 1 shows agreement overall and by report, matching problem CUI and problem CUI with status. Inter-annotator agreement for problem with status was slightly lower for all report types with the largest agreement drop for DS at 15% (11.6 points).

Report Type	CUI	CUI + Status
DS	77.1	65.5
ECHO	83.9	82.8
RAD	84.7	82.8
ECG	89.6	84.8

Table 1: Document-level IAA by report type for problem (CUI) and problem with status (CUI + status)

We report the most common errors by frequency in Table 2. By report type, the most common errors for ECHO, RAD, and ECG were CUI/CUI-less, and DS was Spurious Concepts.

Errors	DS	ECHO	RAD	ECG
SP	423 (42%)	26 (23%)	30 (35%)	8 (18%)
PS	139 (14%)	31 (27%)	8 (9%)	0 (0%)
CS	318 (32%)	9 (8%)	8 (9%)	14 (32%)
CC	110 (11%)	34 (30%)	37 (44%)	22 (50%)
Other	6 (>1%)	14 (13%)	2 (2%)	0 (0%)

Table 2: Error types by frequency - Spurious Problems (SP), Problem Specificity (PS), Conflicting status (CS), CUI/CUI-less (CC)

3.2 Coverage Study

In the training set, there were 203 clinical documents - 93 DS, 37 ECHO, 38 RAD, and 35 ECG. The average number of problems were 22$\pm$10 DS, 10$\pm$4 ECHO, 6$\pm$2 RAD, and 4$\pm$1 ECG. There are 5843 total current problems in SNOMED-CT CORE Problem List. We observed a range of unique SNOMED-CT problem concept frequencies: 776 DS, 63 ECHO, 113 RAD, and 36 ECG by report type. The prevalence of covered problem concepts by CORE is 461 (59%) DS, 36 (57%) ECHO, 71 (63%) RAD, and 16 (44%) ECG. In Table 3, we report coverage of patient problems for each vocabulary. No reports were annotated as "Healthy - no problems". All reports have SNOMED CT coverage of problem mentions above 80%. After mapping problem mentions to CORE, we observed coverage drops for all report types, 24 to 36 points.

Report Type	Patient Problems	Annotated with SNOMED CT	Mapped to CORE
DS	2000	1813 (91%)	1335 (67%)
ECHO	349	300 (86%)	173 (50%)
RAD	190	156 (82%)	110 (58%)
ECG	95	77(81%)	43 (45%)

Table 3: Patient problem coverage by SNOMED-CT and SNOMED-CT CORE

4 Discussion

In this feasibility study, we evaluated how reliably two domain experts can generate a patient problem list and assessed the coverage of annotated patient problems against two standard clinical vocabularies.

4.1 Annotation Study

Overall, we demonstrated that problems can be reliably annotated with moderate to high agreement between domain experts (Table 1). For DS, agreement scores were lowest and dropped most when considering the problem status in the match criteria. The most prevalent disagreement for DS was Spurious problems (Table 2). Spurious problems included additional events (e.g., **C2939181**: *Motor vehicle accident*), procedures (e.g., **C0199470**: *Mechanical ventilation*), and modes of administration (e.g., **C0041281**: *Tube feeding of patient*) that were outside our patient problem list inclusion criteria. Some pertinent findings were also missed. These findings are not surprising given on average more problems occur in DS and the length of DS documents are much longer than other document types. Indeed, annotators are more likely to miss a problem as the number of patient problems increase.

Also, status differences can be attributed to multiple status change descriptions using expressions of time e.g., "cough improved then" and modality "rule out pneumonia", which are harder to track and interpret over a longer document. The most prevalent disagreements for all other document types were CUI/CUI-less in which identifying a CUI representative of a clinical observation proved more difficult. An example of Other disagreement was a sidedness mismatch or redundant patient problem annotation. For example, **C0344911**: *Left ventricular dilatation* vs. **C0344893**: *Right ventricular dilatation* or **C0032285**: *Pneumonia* was recorded twice.

4.2 Coverage Study

We observed that DS and RAD reports have higher counts and coverage of unique patient problem concepts. We suspect this might be because other document types like ECG reports are more likely to have laboratory observations, which may be less prevalent findings in CORE. Across document types, coverage of patient problems in the corpus by SNOMED CT were high ranging from 81% to 91% (Table 3). However, coverage of patient problems by CORE dropped to moderate coverages ranging from 45% to 67%. This suggests that the CORE Problem List is more restrictive and may not be as useful for capturing patient problems from these document types. A similar report of moderate problem coverage with a more restrictive concept list was also reported by Meystre and Haug (2005).

5 Limitations

Our study has limitations. We did not apply a traditional adjudication review between domain experts. In addition, we selected the ShARe corpus from an ICU database in which vocabulary coverage of patient problems could be very different for other domains and specialties.

6 Conclusion

Based on this feasibility study, we conclude that we can generate a reliable patient problem list reference standard for the ShARe corpus and SNOMED CT provides better coverage of patient problems than the CORE Problem List. In future work, we plan to evaluate from each ShARe report type, how well these patient problem lists can be derived and visualized from the individual disease/disorder problem mentions leveraging temporality and modality attributes using natural language processing and machine learning approaches.

Acknowledgments

This work was partially funded by NLM (5T15LM007059 and 1R01LM010964), ShARe (R01GM090187), Swedish Research Council (350-2012-6658), and Swedish Fulbright Commission.

References

Center for Medicare and Medicaid Services. 2013. EHR Incentive Programs-Maintain Problem List. http://www.cms.gov/Regulations-and-Guidance/Legislation/EHRIncentivePrograms/downloads/3_Maintain_Problem_ListEP.pdf.

Noemie Elhadad, Wendy Chapman, Tim OGorman, Martha Palmer, and Guergana. Under Review Savova. under review. The ShARe Schema for the Syntactic and Semantic Annotation of Clinical Texts.

George Hripcsak and Adam S. Rothschild. 2005. Agreement, the F-measure, and Reliability in Information Retrieval. *J Am Med Inform Assoc*, 12(3):296–298.

Stephane Meystre and Peter Haug. 2005. Automation of a Problem List using Natural Language Processing. *BMC Medical Informatics and Decision Making*, 5(30).

Stephane M. Meystre and Peter J. Haug. 2008. Randomized Controlled Trial of an Automated Problem List with Improved Sensitivity. *International Journal of Medical Informatics*, 77:602–12.

Danielle L. Mowery, Pamela W. Jordan, Janyce M. Wiebe, Henk Harkema, John Dowling, and Wendy W. Chapman. 2013. Semantic Annotation of Clinical Events for Generating a Problem List. In *AMIA Annu Symp Proc*, pages 1032–1041.

National Library of Medicine. 2009. The CORE Problem List Subset of SNOMED-CT. Unified Medical Language System 2011. http://www.nlm.nih.gov/research/umls/SNOMED-CT/core_subset.html.

Mohammed Saeed, C. Lieu, G. Raber, and Roger G. Mark. 2002. MIMIC II: a massive temporal ICU patient database to support research in intelligent patient monitoring. *Comput Cardiol*, 29.

Imre Solti, Barry Aaronson, Grant Fletcher, Magdolna Solti, John H. Gennari, Melissa Cooper, and Thomas Payne. 2008. Building an Automated Problem List based on Natural Language Processing: Lessons Learned in the Early Phase of Development. pages 687–691.

Brett R. South, Shuying Shen, Jianwei Leng, Tyler B. Forbush, Scott L. DuVall, and Wendy W. Chapman. 2012. A prototype tool set to support machine-assisted annotation. In *Proceedings of the 2012 Workshop on Biomedical Natural Language Processing*, BioNLP '12, pages 130–139. Association for Computational Linguistics.

Lawrence Weed. 1970. *Medical Records, Medical Education and Patient Care: The Problem-Oriented Record as a Basic Tool.* Medical Publishers: Press of Case Western Reserve University, Cleveland: Year Book.

A System for Predicting ICD-10-PCS Codes from Electronic Health Records

Michael Subotin
3M Health Information Systems
Silver Spring, MD
msubotin@mmm.com

Anthony R. Davis
3M Health Information Systems
Silver Spring, MD
adavis4@mmm.com

Abstract

Medical coding is a process of classifying health records according to standard code sets representing procedures and diagnoses. It is an integral part of health care in the U.S., and the high costs it incurs have prompted adoption of natural language processing techniques for automatic generation of these codes from the clinical narrative contained in electronic health records. The need for effective auto-coding methods becomes even greater with the impending adoption of ICD-10, a code inventory of greater complexity than the currently used code sets. This paper presents a system that predicts ICD-10 procedure codes from the clinical narrative using several levels of abstraction. First, partial hierarchical classification is used to identify potentially relevant concepts and codes. Then, for each of these concepts we estimate the confidence that it appears in a procedure code for that document. Finally, confidence values for the candidate codes are estimated using features derived from concept confidence scores. The concept models can be trained on data with ICD-9 codes to supplement sparse ICD-10 training resources. Evaluation on held-out data shows promising results.

1 Introduction

In many countries reimbursement rules for health care services stipulate that the patient encounter must be assigned codes representing diagnoses that were made for and procedures that were performed on the patient. These codes may be assigned by general health care personnel or by specially trained medical coders. The billing codes used in the U.S. include International Statistical Classification of Diseases and Related Health Problems (ICD) codes, whose version 9 is currently in use and whose version 10 was scheduled for adoption in October 2014[1], as well as Current Procedural Terminology (CPT) codes. The same codes are also used for research, internal bookkeeping, and other purposes.

Assigning codes to clinical documentation often requires extensive technical training and involves substantial labor costs. This, together with increasing prominence of electronic health records (EHRs), has prompted development and adoption of NLP algorithms that support the coding workflow by automatically inferring appropriate codes from the clinical narrative and other information contained in the EHR (Chute et al., 1994; Heinze et al., 2001; Resnik et al., 2006; Pakhomov et al., 2006; Benson, 2006). The need for effective auto-coding methods becomes especially acute with the introduction of ICD-10 and the associated increase of training and labor costs for manual coding.

The novelty and complexity of ICD-10 presents unprecedented challenges for developers of rule-based auto-coding software. Thus, while ICD-9 contains 3882 codes for procedures, the number of codes defined by the ICD-10 Procedure Coding System (PCS) is greater than 70,000. Furthermore, the organization of ICD-10-PCS is fundamentally different from ICD-9, which means that the investment of time and money that had gone into writing auto-coding rules for ICD-9 procedure codes cannot be easily leveraged in the transition to ICD-10.

In turn, statistical auto-coding methods are constrained by the scarcity of available training data with manually assigned ICD-10 codes. While this problem will be attenuated over the years as ICD-10-coded data are accumulated, the health care

[1]The deadline was delayed by at least a year while this paper was in review.

Proceedings of the 2014 Workshop on Biomedical Natural Language Processing (BioNLP 2014), pages 59–67,
Baltimore, Maryland USA, June 26-27 2014.

industry needs effective technology for ICD-10 computer-assisted coding in advance of the implementation deadline. Thus, for developers of statistical auto-coding algorithms two desiderata come to the fore: these algorithms should take advantage of all available training data, including documents supplied only with ICD-9 codes, and they should possess high capacity for statistical generalization in order to maximize the benefits of training material with ICD-10 codes.

The auto-coding system described here seeks to meet both these requirements. Rather than predicting codes directly from the clinical narrative, a set of classifiers is first applied to identify coding-related concepts that appear in the EHR. We use General Equivalence Mappings (GEMs) between ICD-9 and ICD-10 codes (CMS, 2014) to train these models not only on data with human-assigned ICD-10 codes, but also on ICD-9-coded data. We then use the predicted concepts to derive features for a model that estimates probability of ICD-10 codes. Besides the intermediate abstraction to concepts, the code confidence model itself is also designed so as to counteract sparsity of the training data. Rather than train a separate classifier for each code, we use a single model whose features can generalize beyond individual codes. Partial hierarchical classification is used for greater run-time efficiency. To our knowledge, this is the first research publication describing an auto-coding system for ICD-10-PCS. It is currently deployed, in tandem with other auto-coding modules, to support computer-assisted coding in the 3M™360 Encompass™System.

The rest of the paper is organized as follows. Section 2 reviews the overall organization of ICD-10-PCS. Section 4.1 outlines the run-time processing flow of the system to show how its components fit together. Section 4.2 describes the concept confidence models, including the hierarchical classification components. Section 4.3 discusses how data with manually assigned ICD-9 codes is used to train some of the concept confidence models. Section 4.4 describes the code confidence model. Finally, Section 5 reports experimental results.

2 ICD-10 Procedure Coding System

ICD-10-PCS is a set of codes for medical procedures, developed by 3M Health Information Systems under contract to the Center for Medicare and Medicaid Services of the U.S. government. ICD-10-PCS has been designed systematically; each code consists of seven characters, and the character in each of these positions signifies one particular aspect of the code. The first character designates the "section" of ICD-10-PCS: 0 for Medical and Surgical, 1 for Obstetrics, 2 for Placement, and so on. Within each section, the seven components, or axes of classification, are intended to have a consistent meaning; for example in the Medical and Surgical section, the second character designates the body system involved, the third the root operation, and so on (see Table 1 for a list). All procedures in this section are thus classified along these axes. For instance, in a code such as *0DBJ3ZZ*, the *D* in the second position indicates that the body system involved is the gastrointestinal system, *B* in the third position always indicates that the root operation is an excision of a body part, the *J* in the fourth position indicates that the appendix is the body part involved, and the *3* in the fifth position indicates that the approach is percutaneous. The value *Z* in the last two axes means than neither a device nor a qualifier are specified.

Character	Meaning
1st	Section
2nd	Body System
3rd	Root Operation
4th	Body Part
5th	Approach
6th	Device
7th	Qualifier

Table 1: Character Specification of the Medical and Surgical Section of ICD-10-PCS

Several consequences of the compositional structure of ICD-10-PCS are especially relevant for statistical auto-coding methods.

On the one hand, it defines over 70,000 codes, many of which are logically possible, but very rare in practice. Thus, attempts to predict the codes as unitary entities are bound to suffer from data sparsity problems even with a large training corpus. Furthermore, some of the axis values are formulated in ways that are different from how the corresponding concepts would normally be expressed in a clinical narrative. For example, ICD-10-PCS uses multiple axes (root opreration, body part, and, in a sense, the first two axes as well) to encode what many traditional procedure terms (such as those ending in *-tomy* and *-plasty*) express by a single word, while the device axis uses generic categories where a clinical narrative would refer only to specific brand names. This drastically limits how much can be accomplished by matching code descriptions or indexes derived from them against the text of EHRs.

On the other hand, the systematic conceptual structure of PCS codes and of the codeset as a whole can be exploited to compensate for data sparsity and idiosyncracies of axis definitions by introducing abstraction into the model.

3 Related work

There exists a large literature on automatic classification of clinical text (Stanfill et al., 2010). A sizeable portion of it is devoted to detecting categories corresponding to billing codes, but most of these studies are limited to one or a handful of categories. This is in part because the use of patient records is subject to strict regulation. Thus, the corpus used for most auto-coding research up to date consists of about two thousand documents annotated with 45 ICD-9 codes (Pestian et al., 2007). It was used in a shared task at the 2007 BioNLP workshop and gave rise to papers studying a variety of rule-based and statistical methods, which are too numerous to list here.

We limit our attention to a smaller set of research publications describing identification of an entire set of billing codes, or a significant portion thereof, which better reflects the role of auto-coding in real-life applications. Mayo Clinic was among the earliest adopters of auto-coding (Chute et al., 1994), where it was deployed to assign codes from a customized and greatly expanded version of ICD-8, consisting of almost 30K diagnostic codes. A recently reported version of their system (Pakhomov et al., 2006) leverages a combination of example-based techniques and Naïve Bayes classification over a database of over 20M EHRs. The phrases representing the diagnoses have to be itemized as a list beforehand. In another pioneering study, Larkey & Croft (1995) investigated k-Nearest Neighbor, Naïve Bayes, and relevance feedback on a set of 12K discharge summaries, predicting ICD-9 codes. Heinze et al (2000) and Ribeiro-Neto et al (2001) describe systems centered on symbolic computation. Jiang et al (2006) discuss confidence assessment for ICD-9 and CPT codes, performed separately from code generation. Medori & Fairon (2010) combine information extraction with a Naïve Bayes classifier, working with a corpus of about 20K discharge summaries in French. In a recent paper, Perotte et al (2014) study standard and hierarchical classification using support vector machines on a corpus of about 20K EHRs with ICD-9 codes.

We are not aware of any previous publications on auto-coding for ICD-10-PCS, and the results of these studies cannot be directly compared with those reported below due to the unique nature of this code set. Our original contributions also include explicit modeling of concepts and the capability to assign previously unobserved codes within a machine learning framework.

4 Methods

4.1 Run-time processing flow

We first describe the basic run-time processing flow of the system, shown in Figure 1.

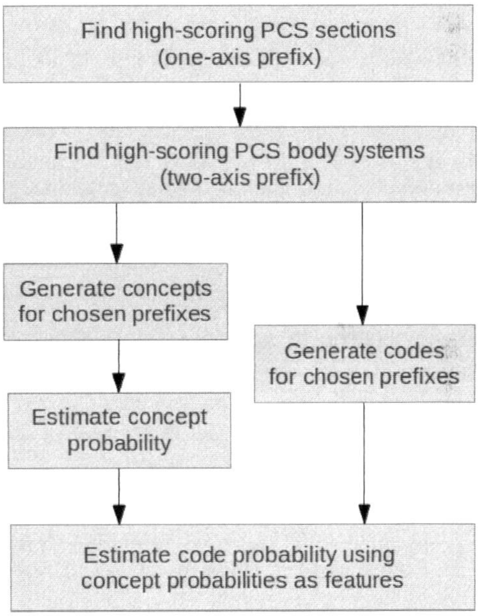

Figure 1: Run-time processing flow

In a naïve approach, one could generate all codes from the ICD-10-PCS inventory for each EHR[2] and estimate their probability in turn, but this would be too computationally expensive. Instead, the hypothesis space is restricted by two-

[2]We use the term EHR generically in this paper. The system can be applied at the level of individual clinical documents or entire patient encounters, whichever is appropriate for the given application.

level hierarchical classification with beam search. First, a set of classifiers estimates the confidence of all PCS sections (one-character prefixes of the codes), one per section. The sections whose confidence exceeds a threshold are used to generate candidate body systems (two-character code prefixes), whose confidence is estimated by another set of classifiers. Then, body systems whose confidence exceeds a threshold are used to generate a set of candidate codes and the set of concepts expressed by these codes. The probability of observing each of the candidate concepts in the EHR is estimated by a separate classifier. Finally, these concept confidence scores are used to derive features for a model that estimates the probability of observing each of the candidate codes, and the highest-scoring codes are chosen according to a thresholding decision rule.

The choice of two hierarchical layers is partially determined by the amount of training data with ICD-10 codes available for this study, since many three-character code prefixes are too infrequent to train reliable classifiers. Given more training data, additional hierarchical classification layers could be used, which would trade a higher risk of recall errors against greater processing speed. The same trade-off can be negotiated by adjusting the beam search threshold.

4.2 Concept confidence models

Estimation of concept confidence – including the confidence of code prefixes in the two hierarchical classification layers – is performed by a set of classifiers, one per concept, which are trained on EHRs supplied with ICD-10 and ICD-9 procedure codes.

The basis for training the concept models is provided by a mapping between codes and concepts expressed by the codes. For example, the code *0GB24ZZ* (Excision of Left Adrenal Gland, Percutaneous Endoscopic Approach) expresses, among other concepts, the concept *adrenal gland* and the more specific concept *left adrenal gland*. It also expresses the concept of *adrenalectomy* (surgical removal of one or both of the adrenal glands), which corresponds to the regular expression *0G[BT][234]..Z* over ICD-10-PCS codes. We used the code-to-concept mapping described in Mills (2013), supplemented by some additional categories that do not correspond to traditional clinical concepts. For example, our set of concepts included entries for the categories of *no device* and *no qualifer*, which are widely used in ICD-10-PCS. We also added entries that specified the device axis or the qualifier axis together with the first three axes, where they were absent in the original concept map, reasoning that the language used to express the choice of the device or qualifier can be specific to particular procedures and body parts.

For data with ICD-10-PCS codes, the logic used to generate training instances is straightforward. Whenever a manually assigned code expresses a given concept, a positive training instance for the corresponding classifier is generated. Negative training instances are sub-sampled from the concepts generated by hierarchical classification layers for that EHR. As can be seen from this logic, the precise question that the concept models seek to answer is as follows: given that this particular concept has been generated by the upstream hierarchical layers, how likely is it that it will be expressed by one of the ICD-10 procedure codes assigned to that EHR?

In estimating concept confidence we do not attempt to localize where in the clinical narrative the given concept is expressed. Our baseline feature set is simply a bag of tokens. We also experimented with other feature types, including frequency-based weighting schemes for token feature values and features based on string matches of Unified Medical Language System (UMLS) concept dictionaries. For the concepts of *left* and *right* we define an additional feature type, indicating whether the token *left* or *right* appears more frequently in the EHR. While still rudimentary, this feature type is more apt to infer laterality than a bag of tokens.

A number of statistical methods can be used to estimate concept confidence. We use the Mallet (McCallum, 2002) implementation of ℓ_1-regularized logistic regression, which has shown good performance for NLP tasks in terms of accuracy as well as scalability at training and run-time (Gao et al., 2007).

4.3 Training on ICD-9 data

In training concept confidence models on data with ICD-9 codes we make use of the General Equivalence Mappings (GEMs), a publicly available resource establishing relationships between ICD-9 and ICD-10 codes (CMS, 2014). Most correspondences between ICD-9 and ICD-10 procedure codes are one-to-many, although other mapping patterns are also found. Furthermore, a code in one set can correspond to a combination of codes from the other set. For example, the ICD-9 code for combined heart-lung transplantation maps to a set of pairs of ICD-10 codes, the first code in the pair representing one of three possible types of heart transplantation, and the other representing one of three possible types of bilateral lung transplantation.

A complete description of the rules underlying GEMs and our logic for processing them is beyond the scope of this paper, and we limit our discussion to the principles underlying our approach. We first distribute a unit probability mass over the ICD-10 codes or code combinations mapped to each ICD-9 code, using logic that reflects the structure of GEMs and distributing probability mass uniformly among comparable alternatives. From these probabilities we compute a cumulative probability mass for each concept appearing in the ICD-10 codes. For example, if an ICD-9 code maps to four ICD-10 codes over which we distribute a uniform probability distibution, and a given concept appears in two of them, we assign the probability of 0.5 to that concept. For a given EHR, we assign to each concept the highest probability it receives from any of the codes observed for the EHR. Finally, we use the resulting concept probabilities to weight positive training instances. Negative instances still have unit weights, since they correspond to concepts that can be unequivocably ruled out based on the GEMs.

4.4 Code confidence model

The code confidence model produces a confidence score for candidate codes generated by the hierarchical classification layers, using features derived from the output of the code confidence models described above. The code confidence model is trained on data with ICD-10 codes. Whenever a candidate code matches a code assigned by human annotators, a positive training instance is generated. Otherwise, a negative instance is generated, with sub-sampling. We report experiments using logistic regression with ℓ_1 and ℓ_2 regularization (Gao et al., 2007).

The definition of features used in the model requires careful attention, because it is in the form of the feature space that the proposed model differs from a standard one-vs-all approach. To elucidate the contrast we may start with a form of the feature space that would correspond to one-vs-all classification. This can be achieved by specifying the identity of a particular code in all feature names. Then, the objective function for logistic regression would decompose into independent learning subproblems, one for each code, producing a collection of one-vs-all classifiers. There are clear drawbacks to this approach. If all parameters are restricted to a specific code, the training data would be fragmented along the same lines. Thus, even if features derived from concepts may seem to enable generalization, in reality they would in each case be estimated only from training instances corresponding to a single code, causing unnecessary data sparsity.

This shortcoming can be overcome in logistic regression simply by introducing generalized features, without changing the rest of the model (Subotin, 2011). Thus, in deriving features from scores of concept confidence models we include only those concepts which are expressed by the given code, but we do not specify the identity of the code in the feature names. In this way the weights for these features are estimated at once from training instances for all codes in which these concepts appear. We combine these generalized features with the code-bound features described earlier. The latter should help us learn more specific predictors for particular procedures, when such predictors exist in the feature space.

While the scores of concept confidence models provide the basis for the feature space of the code confidence model, there are multiple ways in which features can be derived from these scores. The simplest way is to take concept identity (optionally specified by code identity) as the feature name and the confidence score as the feature value. We supplement these features with features based on score quantization. That is, we threshold each concept confidence score at several points and define binary features indicating whether the score exceeds each of the thresholds. For both these feature types, we generate separate features for predictions of concept models trained on ICD-9 data and concept models trained on ICD-10 data in order to allow the code confidence model to learn how useful predictions of concept confidence models are, depending on the type of their training data.

Both the concept confidence models and the

code confidence model can be trained on data with ICD-10 codes. We are thus faced with the question of how best to use this limited resource. The simplest approach would be to train both types of models on all available training data, but there is a concern that predictions of the concept models on their own training data would not reflect their out-of-sample performance, and this would mislead the code confidence model into relying on them too much. An alternative approach, often called stacked generalization (Wolpert, 1992), would be to generate training data for the code confidence model by running concept confidence models on out-of-sample data. We compare the performance of these approaches below.

5 Evaluation

5.1 Methodology

We evaluated the proposed model using a corpus of 28,536 EHRs (individual clinical records), compiled to represent a wide variety of clinical contexts and supplied with ICD-10-PCS codes by trained medical coders. The corpus was annotated under the auspices of 3M Health Information Systems for the express purpose of developing autocoding technology for ICD-10. There was a total of 51,082 PCS codes and 5,650 unique PCS codes in the corpus, only 76 of which appeared in more than 100 EHRs, and 2,609 of which appeared just once. Multiple coders worked on some of the documents, but they were allowed to collaborate, producing what was effectively a single set of codes for each EHR. We held out about a thousand EHRs for development testing and evaluation, each, using the rest for training. The same corpus, as well as 175,798 outpatient surgery EHRs with ICD-9 procedure codes submitted for billing by a health provider were also used to train hierarchical and concept confidence models.

We evaluated auto-coding performance by a modified version of mean reciprocal rank (MRR). MRR is a common evaluation metric for systems with ranked outputs. For a set of Q correct outputs with ranks $rank_i$ among all outputs, standard MRR is computed as:

$$\text{MRR} = \frac{1}{Q}\sum_{i=1}^{Q}\frac{1}{\text{rank}_i}$$

For example, a MRR value of 0.25 means that that the correct answer has rank 4 on average. This metric is designed for tasks where only one of the outputs can be correct. When applied directly to tasks where more than one output can be correct, MRR unfairly penalizes cases with multiple correct outputs, increasing the rank of some correct outputs on account of other, higher-ranked outputs that are also correct. We modify MRR for our task by ignoring correct outputs in the rank computations. In other words, the rank of a correct output is computed as the number of higher-ranked incorrect outputs, plus one. This metric has the advantage of summarizing the accuracy of an auto-coder without reference to a particular choice of threshold, which may be determined by business rules or research considerations, as would be the case for precision and recall.

One advantage of regularized logistic regression is that the value of 1 is often a near-optimal setting for the regularization trade-off parameter. This can save considerable computation time that would be required for tuning this parameter for each experimental condition. We have previously observed that the value of 1 consistently produced near-optimal results for the ℓ_1 regularizer in concept confidence models and for the ℓ_2 regularizer in the code confidence models, and we have used this setting for all the experiments reported here. For the code confidence model with ℓ_1-regularized logistic regression we saw a slight improvement with weaker regularization, and we report the best result we obtained for this model below.

5.2 Results

The results are shown in Table 2. The top MMR score of 0.572 corresponds to a micro-averaged F-score of 0.485 (0.490 precision, 0.480 recall) when the threshold is chosen to obtain approximately equal values for recall and precision[3]. The best result was obtained when:

- the concept models used bag-of-tokens features (with the additional laterality features described in Section 4.2);
- both concept models trained on ICD-9 data and those trained on ICD-10 data were used;
- the code confidence model was trained on data with predictions of concept models trained on all of ICD-10 data (i.e., no

[3]To put these numbers into perspective, note that the average accuracy of trained medical coders for ICD-10 has been estimated to be 63% (HIMSS/WEDI, 2013).

data splitting for stacked generalization was used);

- the code confidence model used all of the feature types described in Section 4.4;
- the code confidence model used logistic regression with ℓ_2 regularization.

We examine the impact of all these choices on system performance in turn.

Model	MRR
All data, all features, ℓ_2 reg.	**0.572**
Concept model training:	
Trained on ICD-10 only	0.558
Trained on ICD-9 only	0.341
Code model features:	
One-vs-all	0.519
No code-bound features	0.553
No quantization features	0.560
Stacked generalization:	
half & half data split	0.501
5-fold cross-validation	0.539
Code model algorithm:	
ℓ_1 regularization	0.528

Table 2: Evaluation results. Each row after the first correponds to varying one aspect of the model shown in the first row. See Section 5.3 for details of the experimental conditions.

5.3 Discussion

Despite its apparent primitive nature, the bag-of-token feature space for the concept confidence models has turned out to provide a remarkably strong baseline. Our experiments with frequency-based weighting schemes for the feature values and with features derived from text matches from the UMLS concept dictionaries did not yield substantial improvements in the results. Thus, the use of UMLS-based features, obtained using Apache ConceptMapper, yielded a relative improvement of 0.6% (i.e., 0.003 in absolute terms), but at the cost of nearly doubling run-time processing time. Nonetheless, we remain optimistic that more sophisticated features can benefit performance of the concept models while maintaining their scalability.

As can be seen from the table, both concept models trained on ICD-9 data and those trained on ICD-10 data contributed to the overall effectiveness of the system. However, the contribution of the latter is markedly stronger. This suggests that further research is needed in finding the best ways of exploiting ICD-9-coded data for ICD-10 autocoding. Given that data with ICD-9 codes is likely to be more readily available than ICD-10 training data in the foreseeable future, this line of investigation holds potential for significant gains in autocoding performance.

For the choice of features used in the code confidence model, the most prominent contribution is made by the feature that generalize beyond specific codes, as discussed in Section 4.4. Adding these features yields a 10% relative improvement over the set of features equivalent to a one-vs-all model. In fact, using the generalized features alone (see the row marked "no code-bound features" in Table 2) gives a score only 0.02 lower than the best result. As would be expected, generalized features are particularly important for codes with limited training data. Thus, if we restrict our attention to codes with fewer than 25 training instances (which account for 95% of the unique codes in our ICD-10 training data), we find that generalized features yielded a 25% relative improvement over the one-vs-all model (0.247 to 0.309). In contrast, for codes with over 100 training instances (which account for 1% of the unique codes, but 36% of the total code volume in our corpus) the relative improvement from generalized features is less than 4% (0.843 to 0.876). These numbers afford two further observations. First, the model can be improved dramatically by adding a few dozen EHRs per code to the training corpus. Secondly, there is still much room for research in mitigating the effects of data sparsity and improving prediction accuracy for less common codes. Elsewhere in Table 2 we see that quantization-based features contribute a modest predictive value.

Perhaps the most surprising result of the series came from investigating the options for using the available ICD-10 training data, which act as training material both for concept confidence models and the code confidence model. The danger of training both type of models on the same corpus is intuitively apparent. If the training instances for the code model are generated by concept models whose training data included the same EHRs, the accuracy of these concept predictions may not reflect out-of-sample performance of the concept models, causing the code model to rely on them excessively.

The simplest implementation of Wolpert's stacked generalization proposal, which is intended to guard against this risk, is to use one part of the corpus to train one predictive layer and use its predictions on the another part of the corpus to train the other layer. The result in Table 2 (see the row marked "half & half data split") shows that the resulting increase in sparsity of the training data for both models leads to a major degradation of the system's performance, even though at runtime concept models trained on all available data are used. We also investigated a cross-validation version of stacked generalization designed to mitigate against this fragmentation of training data. We trained a separate set of concept models on the training portion of each cross-validation fold, and ran them on the held-out portion. The training set for the code confidence model was then obtained by combining these held-out portions. At runtime, concept models trained on all of the available data were used. However, as intuitively compelling as the arguments motivating this procedure may be, the results were not competitive with the baseline approach of using all available training data for all the models.

Finally, we found that an ℓ_2 regularizer performed clearly better than an ℓ_1 regularizer for the code confidence model, even though we set the ℓ_2 trade-off constant to 1 and tuned the ℓ_1 trade-off constant on the development test set. This is in contrast to concept confidence models, where we observed slightly better results with ℓ_1 regularization than with ℓ_2 regularization.

6 Conclusion

We have described a system for predicting ICD-10-PCS codes from the clinical narrative contained in EHRs. The proposed approach seeks to mitigate the sparsity of training data with manually assigned ICD-10-PCS codes in three ways: through an intermediate abstraction to clinical concepts, through the use of data with ICD-9 codes to train concept confidence models, and through the use of a code confidence model whose parameters can generalize beyond individual codes. Our experiments show promising results and point out directions for further research.

Acknowledgments

We would like to thank Ron Mills for providing the crosswalk between ICD-10-PCS codes and clinical concepts; Guoli Wang, Michael Nossal, Kavita Ganesan, Joel Bradley, Edward Johnson, Lyle Schofield, Michael Connor, Jean Stoner and Roxana Safari for helpful discussions relating to this work; and the anonymous reviewers for their constructive criticism.

References

Sean Benson. 2006. Computer-assisted Coding Software Improves Documentation, Coding, Compliance, and Revenue. *Perspectives in Health Information Management, CAC Proceedings*, Fall 2006.

Centers for Medicare & Medicaid Services. 2014. *General Equivalence Mappings. Documentation for Technical Users.* Electronically published at cms.gov.

Chute CG, Yang Y, Buntrock J. 1994. An evaluation of computer assisted clinical classification algorithms. *Proc Annu Symp Comput Appl Med Care.*, 1994:162–6.

Jianfeng Gao, Galen Andrew, Mark Johnson, Kristina Toutanova. 2007. A Comparative Study of Parameter Estimation Methods for Statistical Natural Language Processing. *ACL 2007.*

Daniel T. Heinze, Mark L. Morsch, Ronald E. Sheffer, Jr., Michelle A. Jimmink, Mark A. Jennings, William C. Morris, and Amy E. W. Morsch. 2000. LifeCode^{TM}– A Natural Language Processing System for Medical Coding and Data Mining. *AAAI Proceedings.*

Daniel T. Heinze, Mark Morsch, Ronald Sheffer, Michelle Jimmink, Mark Jennings, William Morris, and Amy Morsch. 2001. LifeCode: A Deployed Application for Automated Medical Coding. *AI Magazine*, Vol 22, No 2.

HIMSS/WEDI. 2013. *ICD-10 National Pilot Program Outcomes Report.* Electronically published at himss.org.

Yuankai Jiang, Michael Nossal, and Philip Resnik. 2006. How Does the System Know It's Right? Automated Confidence Assessment for Compliant Coding. *Perspectives in Health Information Management, Computer Assisted Coding Conference Proceedings*, Fall 2006.

Leah Larkey and W. Bruce Croft. 1995. Automatic Assignment of ICD9 Codes To Discharge Summaries. *Technical report, Center for Intelligent Information Retrieval at University of Massachusetts.*

Andrew Kachites McCallum. 2002. MALLET: A Machine Learning for Language Toolkit. *http://mallet.cs.umass.edu*

Medori, Julia and Fairon, Cédrick. 2010. Machine Learning and Features Selection for Semi-automatic ICD-9-CM Encoding. *Proceedings of the NAACL HLT 2010 Second Louhi Workshop on Text and Data Mining of Health Documents*, 2010: 84–89.

Ronald E. Mills. 2013. Methods using multi-dimensional representations of medical codes. *US Patent Application* US20130006653.

S.V. Pakhomov, J.D. Buntrock, and C.G. Chute. 2006. Automating the assignment of diagnosis codes to patient encounters using example-based and machine learning techniques. *J Am Med Inform Assoc*, 13(5):516–25.

Adler Perotte, Rimma Pivovarov, Karthik Natarajan, Nicole Weiskopf, Frank Wood, Noémie Elhadad . 2014. Diagnosis code assignment: models and evaluation metrics. *J Am Med Inform Assoc*, 21(2):231–7.

Pestian, JP, Brew C, Matykiewicz P, Hovermale DJ, Johnson N, Bretonnel Cohen K, and Duch W. 2007. A shared task involving multi-label classification of clinical free text. *Proceedings ACL: BioNLP*, 2007:97–104.

Philip Resnik, Michael Niv, Michael Nossal, Gregory Schnitzer, Jean Stoner, Andrew Kapit, and Richard Toren. 2006. Using intrinsic and extrinsic metrics to evaluate accuracy and facilitation in computer-assisted coding.. *Perspectives in Health Information Management, Computer Assisted Coding Conference Proceedings*, Fall 2006.

Berthier Ribeiro-Neto, Alberto H.F. Laender and Luciano R.S. de Lima. 2001. An experimental study in automatically categorizing medical documents. *Journal of the American Society for Information Science and Technology*, 52(5): 391–401.

Mary H. Stanfill, Margaret Williams, Susan H. Fenton, Robert A. Jenders, and William R. Hersh. 2010. A systematic literature review of automated clinical coding and classification systems. *J Am Med Inform Assoc.*, 17(6): 646–651.

Michael Subotin. 2011. An exponential translation model for target language morphology. *ACL 2011*.

David H. Wolpert. 1992. Stacked Generalization. *Neural Networks*, 5:241–259.

Structuring Operative Notes using Active Learning

Kirk Roberts[*]
National Library of Medicine
National Institutes of Health
Bethesda, MD 20894
kirk.roberts@nih.gov

Sanda M. Harabagiu
Human Language Technology Research Institute
University of Texas at Dallas
Richardson, TX 75080
sanda@hlt.utdallas.edu

Michael A. Skinner
University of Texas Southwestern Medical Center
Children's Medical Center of Dallas
Dallas, TX 75235
michael.skinner@childrens.com

Abstract

We present an active learning method for placing the event mentions in an operative note into a pre-specified event structure. Event mentions are first classified into action, peripheral action, observation, and report events. The actions are further classified into their appropriate location within the event structure. We examine how utilizing active learning significantly reduces the time needed to completely annotate a corpus of 2,820 appendectomy notes.

1 Introduction

Operative reports are written or dictated after every surgical procedure. They describe the course of the operation as well as any abnormal findings in the surgical process. Template-based and structured methods exist for recording the operative note (DeOrio, 2002), and in many cases have been shown to increase the completeness of surgical information (Park et al., 2010; Gur et al., 2011; Donahoe et al., 2012). The use of natural language, however, is still preferred for its expressive power. This unstructured information is typically the only vehicle for conveying important details of the procedure, including the surgical instruments, incision techniques, and laparoscopic methods employed.

The ability to represent and extract the information found within operative notes would enable powerful post-hoc reasoning methods about surgical procedures. First, the completeness problem may be alleviated by indicating gaps in the surgical narrative. Second, deep semantic similarity methods could be used to discover comparable operations across surgeons and institutions. Third, given information on the typical course and findings of a procedure, abnormal aspects of an operation could be identified and investigated. Finally, other secondary use applications would be enabled to study the most effective instruments and techniques across large amounts of surgical data.

In this paper, we present an initial method for aligning the event mentions within an operative note to the overall event structure for a procedure. A surgeon with experience in a particular procedure first describes the overall event structure. A supervised method enhanced by active learning is then employed to rapidly build an information extraction model to classify event mentions into the event structure. This active learning paradigm allows for rapid prototyping while also taking advantage of the sub-language characteristics of operative notes and the common structure of operative notes reporting the same type of procedure. A further goal of this method is to aid in the evaluation of unsupervised techniques that can automatically discover the event structure solely from the narratives. This would enable all the objectives outlined above for leveraging the unstructured information within operative notes.

This paper presents a first attempt at this active learning paradigm for structuring appendectomy reports. We intentionally chose a well-understood and relatively simple procedure to en-

[*] Most of this work was performed while KR was at the University of Texas at Dallas.

Proceedings of the 2014 Workshop on Biomedical Natural Language Processing (BioNLP 2014), pages 68–76,
Baltimore, Maryland USA, June 26-27 2014.

sure a straight-forward, largely linear event structure where a large amount of data would be easily available. Section 3 describes a generic framework for surgical event structures and the particular structure chosen for appendectomies. Section 4 details the data used in this study. Section 5 describes the active learning experiment for filling in this event structure for operative notes. Section 6 reports the results of this experiment. Section 7 analyzes the method and proposes avenues for further research. First, however, we outline the small amount of previous work in natural language processing on operative notes.

2 Previous Work

An early tool for processing operative notes was proposed by Lamiell et al. (1993). They develop an auditing tool to help enforce completeness in operative notes. A syntactic parser converts sentences in an operative note into a graph structure that can be queried to ensure the necessary surgical elements are present in the narrative. For appendectomies, they could determine whether answers were specified for questions such as "*What was the appendix abnormality?*" and "*Was cautery or drains used?*". Unlike what we propose, they did not attempt to understand the narrative structure of the operative note, only ensure that a small number of important elements were present. Unfortunately, they only tested their rule-based system on four notes, so it is difficult to evaluate the robustness and generalizability of their method.

More recently, Wang et al. (2014) proposed a machine learning (ML) method to extract patient-specific values from operative notes written in Chinese. They specifically extract tumor-related information from patients with hepatic carcinoma, such as the size/location of the tumor, and whether the tumor boundary is clear. In many ways this is similar in purpose to Lamiell et al. (1993) in the sense that there are operation-specific attributes to extract. However, while the auditing function primarily requires knowing whether particular items were stated, their method extracts the particular values for these items. Furthermore, they employ an ML-based conditional random field (CRF) trained and tested on 114 operative notes. The primary difference between the purpose of these two methods and the purpose of our method lies in the attempt to model all the events that characterize a surgery. Both the work of Lamiell et al. (1993) and Wang et al. (2014) can be used for completeness testing, and Wang et al. (2014) can be used to find similar patients. The lack of understanding of the event structure, however, prevents these methods from identifying similar surgical methods or unexpected surgical techniques, or from accomplishing many other secondary use objectives.

In a more similar vein to our own approach, Wang et al. (2012) studies actions (a subset of event mentions) within an operative note. They note that various lexico-syntactic constructions can be used to specify an action (e.g., *incised, the incision was carried, made an incision*). Like our approach, they observed sentences can be categorized into actions, perceptions/reports, and other (though we make this distinction at the event mention level). They adapted the Stanford Parser (Klein and Manning, 2003) with the Specialist Lexicon (Browne et al., 1993) similar to Huang et al. (2005). They do not, however, propose any automatic system for recognizing and categorizing actions. Instead, they concentrate on evaluating existing resources. They find that many resources, such as UMLS (Lindberg et al., 1993) and FrameNet (Baker et al., 1998) have poor coverage of surgical actions, while Specialist and WordNet (Fellbaum, 1998) have good coverage.

A notable limitation of their work is that they only studied actions at the sentence level, looking at the main verb of the independent clause. We have found in our study that multiple actions can occur within a sentence, and we thus study actions at the event mention level. Wang et al. (2012) noted this shortcoming and provide the following illustrative examples:

- *The patient was* **taken** *to the operating room where general anesthesia was* **administered**.
- *After the successful* **induction** *of spinal anesthesia, she was* **placed** *supine on the operating table.*

The second event mention in the first sentence (*administered*) and the first event mention in the second sentence (*induction*) are ignored in Wang et al. (2012)'s study. Despite the fact that they are stated in dependent clauses, these mentions may be more semantically important to the narrative than the mentions in the independent clauses. This is because a grammatical relation does not necessarily imply event prominence. In a further study, Wang et al. (2013) work toward the creation of an automatic extraction system by annotating

PropBank (Palmer et al., 2005) style predicate-argument structures on thirty common surgical actions.

3 Event Structures in Operative Notes

Since operations are considered to be one of the riskier forms of clinical treatment, surgeons follow strict procedures that are highly structured and require significant training and oversight. Thus, a surgeon's description of a particular operation should be highly similar with a different description of the same type of operation, even if written by a different surgeon at a different hospital. For instance, the two examples below were written by two different surgeons to describe the event of controlling the blood supply to the appendix:

- *The 35 mm vascular Endo stapler device was* **fired** *across the mesoappendix...*
- *The meso appendix was* **divided** *with electrocautery...*

In these two examples, the surgeons use different lexical forms (*fired* vs. *divided*), syntactic forms (mesoappendix to the right or left of the EVENT), different semantic predicate-argument structures (INSTRUMENT-EVENT-ANATOMICALOBJECT vs. ANATOMICALOBJECT-EVENT-METHOD), and even different surgical techniques (stapling or cautery). Still, these examples describe the same step in the operation and thus can be mapped to the same location in the event structure.

In order to recognize the event structure in operative notes, we start by specifying an event structure to a particular operation (e.g., mastectomy, appendectomy, heart transplant) and create a ground-truth structure based on expert knowledge. Our goal is then to normalize the event mentions within a operative note to the specific surgical actions in the event structure. While the lexical, syntactic, and predicate-argument structures vary greatly across the surgeons in our data, many event descriptions are highly consistent within notes written by the same surgeon. This is especially true of events with little linguistic variability, typically largely procedural but necessary events that are not the focus of the surgeon's description of the operation. An example of low-variability is the event of placing the patient on the operating table, as opposed to the event of manipulating the appendix to prepare it for removal. Additionally, while there is considerable lexical variation in how an event is mentioned, the terminology for event mentions is fairly limited, resulting in reasonable similarity between surgeons (e.g., the verbal description used for the dividing of the mesoappendix is typically one of the following mentions: *fire*, *staple*, *divide*, *separate*, *remove*).

3.1 Event Structure Representation

Operative notes contain event mentions of many different event classes. Some classes correspond to actions performed by the surgeon, while others describe findings, provide reasonings, or discuss interactions with patients or assistants. These distinctions are necessary to recognizing the event structure of an operation, in which we are primarily concerned with surgical actions. We consider the following event types:

- ACTION: the primary types of events in an operation. These typically involve physical actions taken by the surgeon (e.g., creating/closing an incision, dividing tissue), or procedural events (e.g., anesthesia, transfer to recovery). With limited exceptions, ACTIONs occur in a strict order and the ith ACTION can be interpreted as enabling the $(i+1)^{\text{th}}$ ACTION.
- P_ACTION: the peripheral actions that are optional, do not occur within a specific place in the chain of ACTIONs, and are not considered integral to the event structure. Examples include stopping unexpected bleeding and removing benign cysts un-connected with the operation.
- OBSERVATION: an event that denotes the act of observing a given state. OBSERVATIONs may lead to ACTION (e.g., the appendix is perforated and therefore needs to be removed) or P_ACTIONs (e.g., a cyst is found). They may also be elaborations to provide more details about the surgical method being used.
- REPORT: an event that denotes a verbal interaction between the surgeon and a patient, guardian, or assistant (such as obtaining consent for an operation).

The primary class of events that we are interested in here are ACTIONs. Abstractly, one can view a type of operation as a directed graph with specified start and end states. The nodes denote the events, while the edges denote enablements. An instance of an operation then can be represented as some

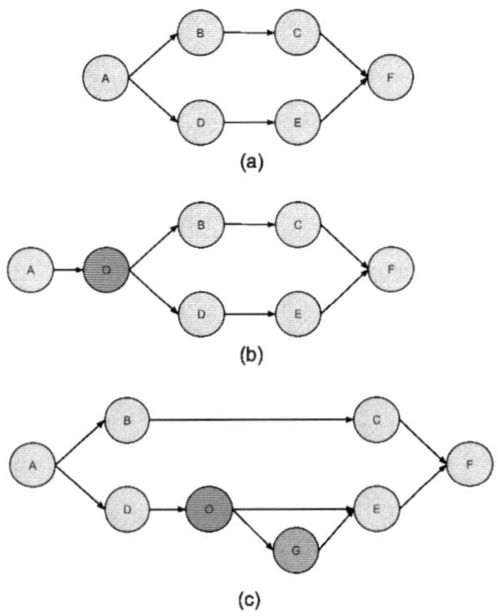

Figure 1: Graphical representation of a surgical procedure with ACTIONS A, B, C, D, E, and F, OBSERVATION O, and P_ACTION G. (a) strict surgical graph (only actions), (b) surgical graph with an observation invoking an action, (c) surgical graph with an observation invoking a peripheral action.

path between the start and end nodes.

In its simplest form, a surgical graph is composed entirely of ACTION nodes (see Figure 1(a)). It is possible to add expected OBSERVATIONS that might trigger a different ACTION path (Figure 1(b)). Finally, P_ACTIONS can be represented as optional nodes in the surgical graph, which may or may not be triggered by OBSERVATIONS (Figure 1(c)). This graphical model is simply a conceptual aid to help design the action types. The model currently plays no role in the automatic classification. For the remainder of this section we focus on a relatively limited surgical procedure that can be interpreted as a linear chain of ACTIONS.

3.2 Appendectomy Representation

Acute appendicitis is a common condition requiring surgical management, and is typically treated by removing the appendix, either laparoscopically or by using an open technique. Appendectomies are the most commonly performed urgent surgical procedure in the United States. The procedure is relatively straight-forward, and the steps of the procedure exhibit little variation between different surgeons. The third author (MS), a surgeon with more than 20 years of experience in pediatric surgery, provided the following primary ACTIONS:

- APP01: transfer patient to operating room
- APP02: place patient on table
- APP03: anesthesia
- APP04: prep
- APP05: drape
- APP06: umbilical incision
- APP07: insert camera/telescope
- APP08: insert other working ports
- APP09: identify appendix
- APP10: dissect appendix away from other structures
- APP11: divide blood supply
- APP12: divide appendix from cecum
- APP13: place appendix in a bag
- APP14: remove bag from body
- APP15: close incisions
- APP16: wake up patient
- APP17: transfer patient to post-anesthesia care unit

In the laparoscopic setting, each of these actions is a necessary part of the operation, and most should be recorded in the operative note. Additionally, any number of P_ACTION, OBSERVATION, and REPORT events may be interspersed.

4 Data

In accordance with generally accepted medical practice and to comply with requirements of The Joint Commission, a detailed report of any surgical procedure is placed in the medical record within 24 hours of the procedure. These notes include the preoperative diagnosis, the post-operative diagnosis, the procedure name, names of surgeon(s) and assistants, anesthetic method, operative findings, complications (if any), estimated blood loss, and a detailed report of the conduct of the procedure. To ensure accuracy and completeness, such notes are typically dictated and transcribed shortly after the procedure by the operating surgeon or one of the assistants.

To obtain the procedure notes for this study, The Children's Medical Center (CMC) of Dallas electronic medical record (EMR) was queried for operative notes whose procedure contained the word "appendectomy" (CPT codes 44970, 44950, 44960) for a preoperative diagnosis of "acute appendicitis" (ICD9 codes 541, 540.0, 540.1). At the time of record acquisition, the CMC EMR had been in operation for about 3 years, and 2,820 notes were obtained, having been completed by 12 pediatric surgeons. In this set, there were 2,757

Surgeon	Notes	Events	Words
$surgeon_1$	8	291	2,305
$surgeon_2$	311	16,379	134,748
$surgeon_3$	143	6,897	57,797
$surgeon_4$	400	8,940	62,644
$surgeon_5$	391	15,246	114,684
$surgeon_6$	307	9,880	77,982
$surgeon_7$	397	10,908	74,458
$surgeon_8$	34	2,401	20,391
$surgeon_9$	2	100	973
$surgeon_{10}$	355	9,987	89,085
$surgeon_{11}$	380	14,211	135,215
$surgeon_{12}$	92	2,417	19,364
Total	2,820	97,657	789,646

Table 1: Overview of corpus by surgeon.

laparoscopic appendectomies and 63 open procedures. The records were then processed automatically to remove any identifying information such as names, hospital record numbers, and dates. For the purposes of this investigation, only the surgeon's name and the detailed procedure note were collected for further study. Owing to the complete anonymity of the records, the study received an exemption from the University of Texas Southwestern Medical Center and CMC Institutional Review Boards. Table 1 contains statistics about the distribution of notes by surgeon in our dataset.

5 Active Learning Framework

Active learning is becoming a more and more popular framework for natural language annotation in the biomedical domain (Hahn et al., 2012; Figueroa et al., 2012; Chen et al., 2013a; Chen et al., 2013b). In an active learning setting, instead of performing manual annotation separate from automatic system development, an existing ML classifier is employed to help choose which examples to annotate. Thus, human annotators can focus on examples that would prove difficult for a classifier, which can dramatically reduce overall annotation time. However, active learning is not without pitfalls, notably sampling bias (Dasgupta and Hsu, 2008), re-usability (Tomanek et al., 2007), and class imbalance (Tomanek and Hahn, 2009). In our work, the purpose of utilizing an active learning framework is to produce a fully-annotated corpus of labeled event mentions in as small a period of time as possible. To some extent, the goal of full-annotation alleviates some of the active learning issues discussed above (re-usability and class imbalance), but sampling bias could still lead to significantly longer annotation time.

Our goal is to (1) distinguish event mentions in one of the four classes introduced in Section 3.1 (event type annotation), and (2) further classify actions into their appropriate location in the event structure (on this data, appendectomy type annotation). While most active learning methods are used with the intention of only manually labeling a sub-set of the data, our goal is to annotate every event mention so that we may ultimately evaluate unsupervised techniques on this data. Our active learning experiment thus proceeds in two parallel tracks: (i) a traditional active learning process where the highest-utility unlabeled event mentions are classified by a human annotator, and (ii) a batch annotation process where extremely similar, "easy" examples are annotated in large groups. Due to small intra-surgeon language variation, and relatively small inter-surgeon variation due to the limited terminology, this second process allows us to annotate large numbers of unlabeled examples at a time. The batch labeling largely annotates unlabeled examples that would not be selected by the primary active learning module because they are too similar to the already-labeled examples. After a sufficient amount of time being spent in traditional active learning, the batch labeling is used to annotate until the batches produced are insufficiently similar and/or wrong classifications are made. After a sufficent number of annotations are made with the active learning method, the choice of when to use the active learning or batch annotation method is left to the discretion of the annotator. This back-and-forth is then repeated iteratively until all the examples are annotated.

For both the active learning and batch labeling processes, we use a multi-class support vector machine (SVM) using a simple set of features:

F1. Event mention's lexical form (e.g., *identified*)
F2. Event mention's lemma (*identify*)
F3. Previous words (*3-the*, *2-appendix*, *1-was*)
F4. Next words (*1-and*, *2-found*, *3-to*, *4-be*, *5-ruptured*)
F5. Whether the event is a gerund (*false*)

Features F3 and F4 were constrained to only return words within the sentence.

To sample event mentions for the active learner, we combine several sampling techniques to ensure a diversity of samples to label. This meta-sampler chooses from 4 different samplers with differing probability p:

1. UNIFORM: Choose (uniformly) an unlabeled instance ($p = 0.1$). Formally, let $\mathcal{L}$ be the

set of manually labeled instances. Then, the probability of selecting an event e_i is:

$$P_{\text{U}}(e_i) \propto \delta(e_i \notin \mathcal{L})$$

Where $\delta(x)$ is the delta function that returns 1 if the condition x is true, and 0 otherwise. Thus, an unlabeled event has an equal probability of being selected as every other unlabeled event.

2. JACCARD: Choose an unlabeled instance biased toward those whose word context is least similar to the labeled instances using Jaccard similarity ($p = 0.2$). This sampler promotes diversity to help prevent sampling bias. Let W_i be the words in e_i's sentence. Then the probability of selecting an event with the JACCARD sampler is:

$$P_{\text{J}}(e_i) \propto \delta(e_i \notin \mathcal{L}) \min_{e_j \in \mathcal{L}} \left[\left(1 - \frac{W_i \cap W_j}{W_i \cup W_j} \right)^{\alpha} \right]$$

Here, α is a parameter to give more weight to dissimilar sentences (we set α = 2).

3. CLASSIFIER: Choose an unlabeled instance biased toward those the SVM assigned low confidence values ($p = 0.65$). Formally, let $f_c(e_i)$ be the confidence assigned by the classifier to event e_i. Then, the probability of selecting an event with the CLASSIFIER sampler is:

$$P_{\text{C}}(e_i) \propto \delta(e_i \notin \mathcal{L})(1 - f_c(e_i))$$

The SVM we use provides confidence values largely in the range (-1, 1), but for some very confident examples this value can be larger. We therefore constrain the raw confidence value $f_r(e_i)$ and place it within the range [0, 1] to achieve the modified confidence $f_c(e_i)$ above:

$$f_c(e_i) = \frac{\max(\min(f_r(e_i), 1), -1) + 1}{2}$$

In this way, $f_c(e_i)$ can be guaranteed to be within [0, 1] and can thus be interpreted as a probability.

4. MISCLASSIFIED: Choose (uniformly) a *labeled* instance that the SVM mis-classifies during cross-validation ($p = 0.05$). Let $f(e_i)$ be the classifier's guess and $\mathcal{L}(e_i)$ be the manual label for event e_i. Then the probability of selecting an event is:

$$P_{\text{M}}(e_i) \propto \delta(e_i \in \mathcal{L})\delta(f(e_i) \neq \mathcal{L}(e_i))$$

Event Type	Precision	Recall	F_1
ACTION	0.79	0.90	0.84
NOT_EVENT	0.75	0.82	0.79
OBSERVATION	0.71	0.57	0.63
P_ACTION	0.66	0.40	0.50
REPORT	1.00	0.58	0.73
Active Learning Accuracy: 76.4%			
Batch Annotation Accuracy: 99.5%			

Table 2: Classification results for event types. Except when specified, results are for data annotated using the active learning method, while the batch annotation results include all data.

The first annotation was made using the UNIFORM sampler. For every new annotation, the meta-sampler chooses one of the above sampling methods according to the above p values, and that sampler selects an example to annotate. For each selected sample, it is first assigned an event type. If it is assigned as an ACTION, the annotator further assigns its appropriate action type. The CLASSIFIER and MISCLASSIFIED samplers alternate between the event type and action type classifiers. These four samplers were chosen to balance the traditional active learning approach (CLASSIFIER), while trying to prevent classifier bias (UNIFORM and JACCARD), while also allowing mis-labeled data to be corrected (MISCLASSIFIED). An evaluation of the utility of the individual samplers is beyond the scope of this work.

6 Results

For event type annotation, two annotators single-annotated 1,014 events with one of five event types (ACTION, P_ACTION, OBSERVATION, REPORT, and NOT_EVENT). The classifier's accuracy on this data was 75.9% (see Table 2 for a breakdown by event type). However, the examples were chosen because they were very different from the current labeled set, and thus we would expect them to be more difficult than a random sampling. When one includes the examples annotated using batch labeling, the overall accuracy is 99.5%.

For action type annotation, the same two annotators labeled 626 ACTIONs with one of the 17 action types (APP01–APP17). The classifier's accuracy on this data was again a relatively low 72.2% (see Table 3 for a breakdown by action type). However, again, these examples were expected to be difficult for the classifier. When one includes the examples annotated using batch labeling, the overall accuracy is 99.4%.

Action Type	Precision	Recall	F_1
APP01	0.91	0.77	0.83
APP02	1.00	0.67	0.80
APP03	1.00	0.67	0.80
APP04	0.95	0.95	0.95
APP05	1.00	1.00	1.00
APP06	0.79	0.72	0.76
APP07	0.58	0.58	0.58
APP08	0.65	0.75	0.70
APP09	0.82	0.93	0.87
APP10	0.63	0.73	0.68
APP11	0.50	0.50	0.50
APP12	0.61	0.56	0.58
APP13	0.94	0.94	0.94
APP14	0.71	0.73	0.72
APP15	0.84	0.79	0.82
APP16	0.93	0.81	0.87
APP17	0.84	0.89	0.86
Active Learning Accuracy: 71.4%			
Batch Annotation Accuracy: 99.4%			

Table 3: Classification results for action types.

7 Discussion

The total time allotted for annotation was approximately 12 hours, split between two annotators (the first author and a computer science graduate student). Prior to annotation, both annotators were given a detailed description of an appendectomy, including a video of a procedure to help associate the actual surgical actions with the narrative description. After annotation, 1,042 event types were annotated using the active learning method, 90,335 event types were annotated using the batch method, and 6,279 remained un-annotated. Similarly, 658 action types were annotated using the active learning method, 35,799 action types were annotated using the batch method, and 21,151 remained un-annotated. A greater proportion of actions remained un-annotated due to the lower classifier confidence associated with the task. Event and action types were annotated in unison, but we estimate during the active learning process it took about 25 seconds to annotate each event (both the event type and the action type if classified as an ACTION). The batch process enabled the annotation of an average of 3 event mentions per second.

This rapid annotation was made possible by the repetitive nature of operative notes, especially within an individual surgeon's notes. For example, the following statements were repeated over 100 times in our corpus:

- *General anesthesia was* **induced**.
- *A Foley catheter was* **placed** *under sterile conditions.*
- *The appendix was* **identified** *and seemed to be acutely* **inflamed**.

The first example was used by an individual surgeon in 95% of his/her notes, and only used three times by a different surgeon. In the second example, the sentence is used in 77% of the surgeon's notes while only used once by another surgeon. The phrase "*Foley catheter was placed*", however, was used 133 times by other surgeons. In the context of an appendectomy, this action is unambiguous, and so only a few annotations are needed to recognize the hundreds of actual occurrences in the data. Similarly, with the third example, the phrase "*the appendix was identified*" was used in over 600 operative notes by 10 of the 12 surgeons. After a few manual annotations to achieve sufficient classification confidence, the batch process can identify duplicate or near-duplicate events that can be annotated at once, greatly reducing the time needed to achieve full annotation.

Unfortunately, the most predictable parts of a surgeon's language are typically the least interesting from the perspective of understanding the critical points in the narrative. As shown in the examples above, the highest levels of redundancy are found in the most routine aspects of the operation. The batch annotation, therefore, is quite biased and the 99% accuracies it achieves cannot be expected to hold up once the data is fully annotated. Conversely, the active learning process specifically chooses examples that are different from the current labeled set and thus are more difficult to classify. Active learning is more likely to sample from the "long tail" than the most frequent events and actions, so the performance on the chosen sample is certainly a lower bound on the performance of a completely annotated data set. If one assumes the remaining un-annotated data will be of similar difficulty to the data sampled by the active learner, one could project an overall event type accuracy of 97% and an overall action type accuracy of 89%. This furthermore assumes no improvements are made to the machine learning method based on this completed data.

One way to estimate the potential bias in batch annotation is by observing the differences in the distributions of the two data sets. Figure 2 shows the total numbers of action types for both the active learning and batch annotation portions of the data. For the most part, the distributions are similar. APP08 (insert other working ports), APP10 (dissect appendix away from other structures), APP11 (divide blood supply), APP12 (di-

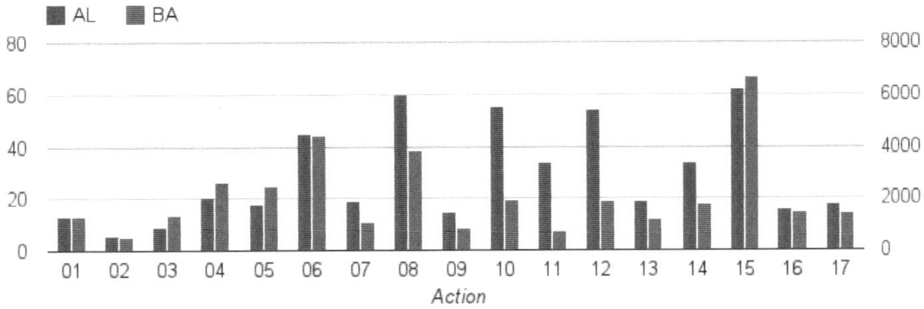

Figure 2: Frequencies of action types in the active learning (AL) portion of the data set (left vertical axis) and the batch annotation (BA) portion of the data set (right vertical axis).

vide appendix from cecum), and APP14 (remove bag from body) are the most under-represented in the batch annotation data. This confirms our hypothesis that some of the most interesting events have the greatest diversity in expression.

In Section 2 we noted that a limitation of the annotation method of Wang et al. (2012) was that a sentence could only have one action. We largely overcame this problem by associating a single surgical action with an event mention. This has one notable limitation, however, as occasionally a single event mention corresponds to more than one action. In our data, APP11 and APP12 are commonly expressed together:

- *Next, the mesoappendix and appendix is* **stapled**$_{\text{APP11/APP12}}$ *and then the appendix is* **placed**$_{\text{APP13}}$ *in an endobag.*

Here, a coordination ("*mesoappendix and appendix*") is used to associate two events (the stapling of the mesoappendix and the stapling of the appendix) with the same event mention. In the event extraction literature, this is a well-understood occurrence, as for instance TimeML (Pustejovsky et al., 2003) can represent more than one event with a single event mention. In practice, however, few automatic TimeML systems handle such phenomena. Despite this, for our purpose the annotation structure should likely be amended so that we can account for all the important actions in the operative note. This way, gaps in our event structure will correspond to actual gaps in the narrative (e.g., dividing the blood supply is a critical step in an appendectomy and therefore needs to fit within the event structure).

Finally, the data in our experiment comes from a relatively simple procedure (an appendectomy). It is unclear how well this method would generalize to more complex operations. Most likely, the difficulty will lie in actions that are highly ambiguous, such as if more than one incision is made. In this case, richer semantic information will be necessary, such as the spatial argument that indicates where a particular event occurs (Roberts et al., 2012).

8 Conclusion

With the increasing availability of electronic operative notes, there is a corresponding need for deep analysis methods to understand the note's narrative structure to enable applications for improving patient care. In this paper, we have presented a method for recognizing how event mentions in an operative note fit into the event structure of the actual operation. We have proposed a generic framework for event structures in surgical notes with a specific event structure for appendectomy operations. We have described a corpus of 2,820 operative notes of appendectomies performed by 12 surgeons at a single institution. With the ultimate goal of fully annotating this data set, which contains almost 100,000 event mentions, we have shown how an active learning method combined with a batch annotation process can quickly annotate the majority of the corpus. The method is not without its weaknesses, however, and further annotation is likely necessary.

Beyond finishing the annotation process, our ultimate goal is to develop unsupervised methods for structuring operative notes. This would enable expanding to new surgical procedures without human intervention while also leveraging the increasing availability of this information. We have shown in this work how operative notes have linguistic characteristics that result in parallel structures. It is our goal to leverage these characteristics in developing unsupervised methods.

Acknowledgments

The authors would like to thank Sanya Peshwani for her help in annotating the data.

References

Collin F. Baker, Charles J. Fillmore, and John B. Lowe. 1998. The Berkeley FrameNet project. In *Proceedings of ACL/COLING*.

Allen C. Browne, Alexa T. McCray, and Suresh Srinivasan. 1993. The SPECIALIST Lexicon. Technical Report NLM-LHC-93-01, National Library of Medicine.

Yukun Chen, Hongxin Cao, Qiaozhu Mei, Kai Zheng, and Hua Xu. 2013a. Applying active learning to supervised word sense disambiguation in MEDLINE. *J Am Med Inform Assoc*, 20:1001–1006.

Yukun Chen, Robert Carroll, Eugenia R. McPeek Hinz, Anushi Shah, Anne E. Eyler, Joshua C. Denny, , and Hua Xu. 2013b. Applying active learning to high-throughput phenotyping algorithms for electronic health records data. *J Am Med Inform Assoc*, 20:e253–e259.

Sanjoy Dasgupta and Daniel Hsu. 2008. Hierarchical Sampling for Active Learning. In *Proceedings of the International Conference on Maching Learning*.

J.K. DeOrio. 2002. Surgical templates for orthopedic operative reports. *Orthopedics*, 25(6):639–642.

Laura Donahoe, Sean Bennett, Walley Temple, Andrea Hilchie-Pye, Kelly Dabbs, Ethel MacIntosh, and Geoff Porter. 2012. Completeness of dictated operative reports in breast cancer–the case for synoptic reporting. *J Surg Oncol*, 106(1):79–83.

Christiane Fellbaum. 1998. *WordNet: An Electronic Lexical Database*. MIT Press.

Rosa L. Figueroa, Qing Zeng-Treitler, Long H. Ngo, Sergey Goryachev, and Eduardo P. Wiechmann. 2012. Active learning for clinical text classification: is it better than random sampling? *J Am Med Inform Assoc*, 19:809–816.

I. Gur, D. Gur, and J.A. Recabaren. 2011. The computerized synoptic operative report: A novel tool in surgical residency education. *Arch Surg*, pages 71–74.

Udo Hahn, Elena Beisswanger, Ekaterina Buyko, and Erik Faessler. 2012. Active Learning-Based Corpus Annotation – The PATHOJEN Experience. In *Proceedings of the AMIA Symposium*, pages 301–310.

Yang Huang, Henry J Lowe, Dan Klein, and Russell J Cucina. 2005. Improved Identification of Noun Phrases in Clinical Radiology Reports Using a High-Performance Statistical Natural Language Parser Augmented with the UMLS Specialist Lexicon. *J Am Med Inform Assoc*, 12:275–285.

Dan Klein and Christopher D. Manning. 2003. Accurate Unlexicalized Parsing. In *Proceedings of ACL*, pages 423–430.

James M Lamiell, Zbigniew M Wojcik, and John Isaacks. 1993. Computer Auditing of Surgical Operative Reports Written in English. In *Proc Annu Symp Comput Appl Med Care*, pages 269–273.

Donald A.B. Lindberg, Betsy L. Humphreys, and Alexa T. McCray. 1993. The Unified Medical Language System. *Methods of Information in Medicine*, 32(4):281–291.

Martha Palmer, Paul Kingsbury, and Daniel Gildea. 2005. The Proposition Bank: An Annotated Corpus of Semantic Roles. *Computational Linguistics*, 31(1):71–106.

Jason Park, Venu G. Pillarisetty, Murray F. Brennan, and *et al.* 2010. Electronic Synoptic Operative Reporting: Assessing the Reliability and Completeness of Synoptic Reports for Pancreatic Resection. *J Am Coll Surgeons*, 211(3):308–315.

James Pustejovsky, José Castano, Robert Ingria, Roser Saurí, Robert Gaizauskas, Andrea Setzer, Graham Katz, and Dragomir Radev. 2003. TimeML: Robust Specification of Event and Temporal Expressions in Text. In *Proceedings of the Fifth International Workshop on Computational Semantics*.

Kirk Roberts, Bryan Rink, Sanda M. Harabagiu, Richard H. Scheuermann, Seth Toomay, Travis Browning, Teresa Bosler, and Ronald Peshock. 2012. A Machine Learning Approach for Identifying Anatomical Locations of Actionable Findings in Radiology Reports. In *Proceedings of the AMIA Symposium*.

Katrin Tomanek and Udo Hahn. 2009. Reducing Class Imbalance during Active Learning for Named Entity Annotation. In *Proceedings of KCAP*.

Katrin Tomanek, Joachim Wermter, and Udo Hahn. 2007. An Approach to Text Corpus Construction which Cuts Annotation Costs and Maintains Reusability of Annotated Data. In *Proceedings of EMNLP/CoNLL*, pages 486–495.

Yan Wang, Serguei Pakhomov, Nora E. Burkart, James O. Ryan, and Genevieve B. Melton. 2012. A Study of Actions in Operative Notes. In *Proceedings of the AMIA Symposium*, pages 1431–1440.

Yan Wang, Serguei Pakhomov, and Genevieve B Melton. 2013. Predicate Argument Structure Frames for Modeling Information in Operative Notes. In *Studies in Health Technology and Informatics (MEDINFO)*, pages 783–787.

Hui Wang, Weide Zhang, Qiang Zeng, Zuofeng Li, Kaiyan Feng, and Lei Liu. 2014. Extracting important information from Chinese Operation Notes with natural language processing methods. *J Biomed Inform*.

Chunking Clinical Text Containing Non-Canonical Language

Aleksandar Savkov
Department of Informatics
University of Sussex
Brighton, UK
a.savkov@sussex.ac.uk

John Carroll
Department of Informatics
University of Sussex
Brighton, UK
j.a.carroll@sussex.ac.uk

Jackie Cassell
Primary Care and Public Health
Brighton and Sussex Medical School
Brighton, UK
j.cassell@bsms.ac.uk

Abstract

Free text notes typed by primary care physicians during patient consultations typically contain highly non-canonical language. Shallow syntactic analysis of free text notes can help to reveal valuable information for the study of disease and treatment. We present an exploratory study into chunking such text using off-the-shelf language processing tools and pre-trained statistical models. We evaluate chunking accuracy with respect to part-of-speech tagging quality, choice of chunk representation, and breadth of context features. Our results indicate that narrow context feature windows give the best results, but that chunk representation and minor differences in tagging quality do not have a significant impact on chunking accuracy.

1 Introduction

Clinical text contains rich, detailed information of great potential use to scientists and health service researchers. However, peculiarities of language use make the text difficult to process, and the presence of sensitive information makes it hard to obtain adequate quantities for developing processing systems. The short term goal of most research in the area is to achieve a reliable language processing foundation that can support more complex tasks such as named entity recognition (NER) to a sufficiently reliable level.

Chunking is the task of identifying non-recursive phrases in text (Abney, 1991). It is a type of shallow parsing that is a less challenging task than dependency or constituency parsing. This makes it likely to give more reliable results on clinical text, since there is a very limited amount of annotated (or even raw) text of this kind available for system development. Even though chunking does not provide as much syntactic information as full parsing, it is an excellent method for identifying base noun phrases (NP), which is a key issue in symptom and disease identification. Identifying symptoms and diseases is at the heart of harnessing the potential of clinical data for medical research purposes.

There are few resources that enable researchers to adapt general domain techniques to clinical text. Using the Harvey Corpus[1] – a chunk annotated clinical text language resource – we present an exploratory study into adapting general domain tools and models to apply to free text notes typed by UK primary care physicians.

2 Related Work

The Mayo Clinic Corpus (Pakhomov et al., 2004) is a key resource that has been widely used as a gold standard in part-of-speech (POS) tagging of clinical text. Based on that corpus and the Penn TreeBank (Marcus et al., 1993), Coden et al. (2005) present an analysis of the effects of domain data on the performance of POS tagging models, demonstrating significant improvements with models trained entirely on domain data. Savova et al. (2010) use this corpus for the development of cTAKES, Mayo Clinic's processing pipeline for clinical text.

Fan et al. (2011) show that using more diverse clinical data can lead to more accurate POS tagging. They report that models trained on clinical text datasets from two different institutions perform on each of the datasets better than both models trained only on the same or the other dataset.

Fan et al. (2013) present guidelines for syntactic parsing of clinical text and a clinical Treebank annotated according to them. The guidelines are designed to help the annotators handle the non-canonical language that is typical of clinical text.

[1] An article describing the corpus is currently under review.

Proceedings of the 2014 Workshop on Biomedical Natural Language Processing (BioNLP 2014), pages 77–82,
Baltimore, Maryland USA, June 26-27 2014.

3 Data

The Harvey Corpus is a chunk-annotated corpus consisting of pairs of manually anonymised UK primary care physician (General Practitioner, or GP) notes and associated Read codes (Bentley et al., 1996). Each Read code has a short textual gloss. The purpose of the codes is to make it easy to extract structured data from clinical records. The reason we include the codes in the corpus is that GPs often use their glosses as the beginning of their note. Two typical examples (without chunk annotation for clarity) are shown below.

***Birth details** || Normal deliviery Girl Weight - 3. 960kg Apgar score @ 1min - 9 Apgar score @ 5min - 9 Vit K given Paed check NAD HC - 34. 9cm Hip test performed* (1)

***Chest pain** || musculoskel pain last w/e, nil to find, ecg by paramedic no change, reassured, rev sos* (2)

The corpus comprises 890 pairs of Read codes and notes, each annotated by medical experts using a chunk annotation scheme that includes non-recursive noun phrases (NPs), main verb groups (MVs), and a common annotation for adjectival and adverbial phrases (APs). Example (3) below illustrates the annotation. The majority of the records (750) were double blind annotated by medical experts, after which the resulting annotation was adjudicated by a third medical expert annotator.

*[**Chest pain**]^{NP} || [musculoskel pain]^{NP} [last w/e]^{NP}, [nil]^{AP} to [find]^{MV}, [ecg]^{NP} by [paramedic]^{NP} [no change]^{NP}, [reassured]^{MV}, [rev]^{MV} [sos]^{AP}* (3)

Inter-annotator agreement was 0.86 f-score, taking one annotator to be the gold standard and the other the candidate. We calculate the f-score according to the MUC-7 (Chinchor, 1998) specification, with the standard f-score formula. The calculation is kept symmetric with regard to the choice of gold standard annotator by limiting the counting of *incorrect* categories to one per tag, and equating the *missing* and *spurious* categories. For example, three words annotated as one three-token chunk by annotator A and three one-token chunks by annotator B will have one incorrect and two missing/spurious elements.

The rest of the records are a by-product of the training process. Ninety records were triple annotated by three different medical experts with the help of a computational linguist, and fifty records were double annotated by a medical expert – alone and together with a computational linguist.

It is important to note that the text in the corpus is not representative of all types of GP notes. It is focused on text that represents the dominant part of day-to-day notes, rather than standard edited text such as copies of letters to specialists and other medical practitioners.

Even though the corpus data is very rich in information, its non-canonical language means that it is very different from other clinical corpora such as the Mayo Clinic Corpus (Pakhomov et al., 2004) and poses different challenges for processing. The GP notes in the Harvey Corpus can be regarded as groups of medical 'tweets' meant to be used mainly by the author. Sentence segmentation in the classical sense of the term is often impossible, because there are no sentences. Instead there are short bursts of phrases concatenated together often without any indication of their boundaries. The average length of a note is roughly 30 tokens including the Read code. This is in contrast to notes in other clinical text datasets, which range from 100 to 400 tokens on average (Fan et al., 2011; Pakhomov et al., 2004). As well as typical clinical text characteristics such as domain-specific acronyms, slang, and abbreviations, punctuation and casing are often misleading (if present at all), and some common classes of words (e.g. auxiliary verbs) are almost completely absent.

4 Chunking

State-of-the-art text chunking accuracy reaches an f-score of 95% (Sun et al., 2008). However, this is for standard, edited text, and relies on accurate POS tagging in a pre-processing step. However, the characteristics of GP-written free text make accurate part of speech (POS) tagging and chunking difficult. Major problems are caused by unknown tokens and ambiguities due to omitted words or phrases.

We evaluate two standard chunking tools, YamCha (Kudo and Matsumoto, 2003) and CRF++[2], selected based on their support for trainable context features. The tools were applied to the Har-

[2] `http://crfpp.googlecode.com/svn/trunk/doc/index.html`

	POS	YamCha IOB	YamCha BEISO	CRF++ IOB	CRF++ BEISO
ARK_{IRC}	75.35	76.63 σ1.04	76.87 σ2.91	75.87 σ1.64	76.23 σ1.99
$ARK_{Twitter}$	–	**76.72 σ2.11**	**77.53 σ1.65**	**76.63 σ2.36**	**77.23 σ1.06**
ARK_{Ritter}	75.70	76.59 σ2.01	76.72 σ2.11	**76.63 σ1.05**	77.17 σ1.77
cTAKES	**82.42**	75.32 σ2.52	75.85 σ2.02	75.43 σ1.79	75.53 σ1.90
GENIA	80.63	71.70 σ2.27*	74.86 σ1.41	74.16 σ2.03*	74.19 σ1.72
RASP	–	74.24 σ1.84	75.10 σ1.31	75.63 σ2.33	75.76 σ2.18
Stanford	80.68	76.40 σ1.69	76.36 σ2.92	75.95 σ1.25	75.94 σ1.91
SVMTool	76.40	74.32 σ2.57	74.30 σ2.71	74.66 σ1.77	74.68 σ2.28
Wapiti	73.39	74.74 σ2.29	74.78 σ1.33	73.59 σ2.62	73.83 σ2.31
baseline	–	69.66 σ1.89*	69.76 σ1.24	67.05 σ1.15*	68.65 σ1.41

Table 1: Chunking results using *YamCha* and *CRF++* on data automatically POS tagged using nine different models; the baseline is with no tagging. The IOB and BEISO columns compare the impact of two chunk representation strategies. The POS column indicates the part-of-speech tagging accuracy for a subset of the corpus. Asterisks indicate pairs of significantly different YamCha and CRF++ results (t-test with 0.05 p-value).

vey Corpus with automatically generated POS annotation. Given the small amount of data and the challenges presented above, we expected that our results would be lower than those reported by Savova et al. (2010). The aim of these experiments is to find the best performance obtainable with standard chunking tools, which we will build on in further stages of our research.

We conducted pairs of experiments, one with each chunking tool, divided into three groups: the first investigates the effects of choice of POS tagger for training data annotation (Section 4.1); the second compares two chunk representations (Section 4.2); and the third searches for the optimal context features (Section 4.3). All feature tuning experiments were conducted on a development set and tested using 10-fold cross-validation on the rest of the data. We used 10% of the whole data for the development set and 90% of the remaining data for a training sample during development. This guarantees the development model is trained on the same amount of data as the testing model.

4.1 Part-of-Speech Tagging

We evaluated and compared the results yielded by the two chunkers, having applied each of seven off-the-shelf POS taggers. Of these taggers, cTAKES (Savova et al., 2010) and GENIA (Tsuruoka et al., 2005) are the only ones trained on data that resembles ours, which suggests that they should have the best chance of performing well. We also selected a number of other taggers while trying to diversify their algorithms and training data as much as possible: the POS tagger part of the Stanford NLP package (Toutanova et al., 2003) because it is one of the most successfully applied in the field; the RASP tagger (Briscoe et al., 2006) because of its British National Corpus (Clear, 1993) training data; the ARK tagger (Owoputi et al., 2013) because of the terseness of the tweet language; and the SVMTool (Giménez and Màrquez, 2004) and Wapiti (Lavergne et al., 2010) because they use SVM and CRF algorithms. Our baseline model uses no part of speech information.

Using the Penn TreeBank tagset (Marcus et al., 1993), we manually annotated a subset of the corpus of comparable size to the development set. Using this dataset we estimated the tagging accuracy for all models that support that tagset (omitting *RASP* and *ARK Twitter* since they use different tagsets). In this evaluation, cTAKES is the best performing model, followed closely by the Stanford POS tagger and GENIA.

The results in Table 1 show that the differences between chunking models trained on different POS annotations are small and mostly not statistically significant from each other. However, all the results are significantly better than the baseline, apart from those based on the GENIA tagger output.

4.2 Chunk Representation

The dominant chunk representation standard *inside, outside, begin* (IOB) introduced by Ramshaw and Marcus (1995) and established with the

CoNLL-2000 shared task (Sang and Buchholz, 2000) takes a minimalistic approach to the representation problem in order to keep the number of labels low. Note that for chunking representations the total number of labels is the product of the chunk types and the set of representation types plus the outside tag, meaning that for IOB with our set of three chunk types (NP, MV, AP) there are seven labels.

Alternative chunk representations, such as *begin, end, inside, single, outside* (BEISO)[3] as used by Kudo and Matsumoto (2001), offer more fine-grained tagsets, presumably at a performance cost. That cost is unnecessary unless there is something to be gained from a more fine-grained tagset at decoding time, because the two representations are deterministically inter-convertible. For instance, an *end* tag could be useful for better recognising boundaries between chunks of the same type. The BEISO tagset model looks for the boundary before and after crossing it, while an IOB model only looks after. This should give only a small gain with standard edited text because the chunk type distribution is fairly well balanced and punctuation divides ambiguous cases such as lists of compound nouns. However, the Harvey Corpus is NP-heavy and contains many sequences of NP chunks that do not have any punctuation to mark their boundaries.

We evaluated the two chunk representations in combination with each POS tagger. Table 1 shows that the differences between the results for the two representations are small and never statistically significant. We also evaluated the two chunk representations with different amounts of training data. The resulting learning curves (Figure 1) are almost identical.

4.3 Context Features

We approached the feature tuning task by first exploring the smaller feature space of YamCha and then using the trends there to constrain the features of CRF++. YamCha has three groups of features responsible for tokens, POS tags and dynamically generated (i.e. preceding) chunk tags. For all experiments we determined the best feature set by exhaustively testing all context feature combinations within a predefined range. We used the same context window for the token and tag features in order to reduce the search space. Given

[3]Also sometimes abbreviated IOBSE

Feature Set	CV	Dev
W_{-1}-W_1, T_{-1}-T_1, C_{-1}	77.28 σ1.9	75.28
W_{-1}-W_1, T_{-1}-T_1, C_{-2}-C_{-1}	77.27 σ2.6	74.70
W_{-1}-W_2, T_{-1}-T_2, C_{-1}	76.86 σ1.5	74.08
W_{-2}-W_1, T_{-2}-T_1, C_{-2}	76.46 σ1.3	74.00
W_{-1}-W_1, T_{-1}-T_1, C_{-2}	76.89 σ2.1	73.92
W_{-2}-W_1, T_{-2}-T_1, C_{-3}-C_{-1}	76.52 σ0.9	73.91
W_{-1}-W_1, T_{-1}-T_1, C_{-3}-C_{-1}	77.02 σ2.0	73.90
W_{-2}-W_2, T_{-2}-T_2, C_{-1}	77.03 σ1.9	73.86
W_{-1}-W_1, T_{-1}-T_1, C_{-3}	77.15 σ1.5	73.63
W_{-3}-W_1, T_{-3}-T_1, C_{-2}-C_{-1}	75.71 σ1.9	73.63

Table 2: Development set and 10-fold cross-validation results for the top ten feature sets of YamCha models trained on $\text{ARK}_{\text{Twitter}}$ POS annotation. Token features are represented with *W*, POS features with *T*, and dynamically generated chunk features with *C*. None of the cross-validation results are significantly different from each other (t-test with 0.05 p-value).

the terseness of the text we expected that wider context windows of more than three tokens would not be beneficial to the model, and therefore did not consider them. Our experiments using YamCha confirmed this hypothesis and showed a consistent trend among all experiments in favouring a window of -1 to +1 for tokens and slightly wider for chunk tags (see Table 2).

CRF++ provides a more powerful feature configuration allowing for unary and pairwise[4] features of output tags. The unary features allow the construction of token or POS tag bigrams and trigrams in addition to the standard context windows. The feature tuning search space with so many parameters is enormous, which required us to use our findings from the YamCha experiments to trim it down and make it computationally feasible. First, we decreased the search window of all features by one in each direction from -3:3 to -2:2. Second, we used the top scoring POS model from the first experimental runs to constrain the features even further by selecting only the top one hundred for the rest of the models.

We could not identify the same uniform trend in the top feature sets as we could with YamCha. Our results ranged from very small context windows to the maximum size of our search space. How-

[4]The unary and pairwise features of output tags are referred to as *unigram and bigram features of output tags* on the CRF++ web page. Although this is correct, it can also be confused with unigrams and bigrams of tokens, which are expressed as unary (unigram) output tag features.

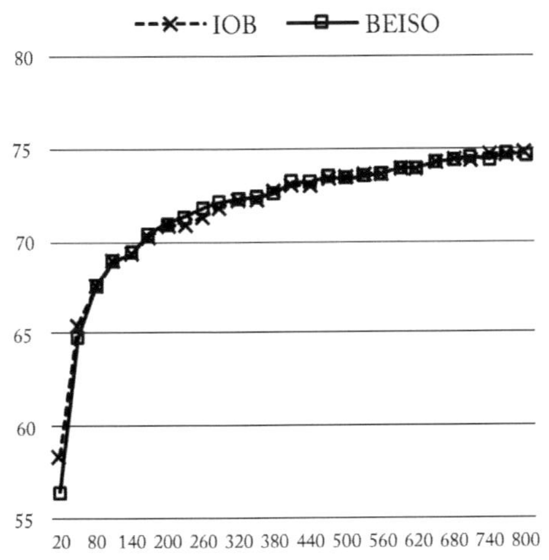

Figure 1: Chunking results for YamCha IOB and BEISO models with increasing amounts of training data.

ever, we noticed that BEISO feature sets tend to be smaller than the IOB ones. We also found that the pairwise features normally improve the results.

5 Discussion and Future Work

We were surprised that the experiments did not show a clear correlation between POS tagging accuracy and chunking accuracy. On the other hand, the chunking results using POS tagged data are significantly better than the baseline, except when using the GENIA tagger output. The small differences between training sets of similar POS accuracy could be explained due to the non-uniform impact of the wrong POS tag on the chunking process. Some mistakes such as labelling a noun as a verb in the middle of a NP chunk are almost sure to propagate and cause further chunking errors, whereas others may have minimal or no effect, for example labelling a singular noun as a proper noun. An error analysis of verb tags and noun tags (Table 3) shows that the ARK models tend to make more mistakes that keep the annotation within the same tag group compared to the GENIA model (see column pairs 1 and 3, and 2 and 4). This is a possible explanation for the lower accuracy of the chunking model trained on data tagged by GENIA.

Our experiments showed that the models using the two chunk representations did not perform significantly differently from each other. We also showed that this conclusion is likely to hold if more training data were available.

Model	N_{group}	V_{group}	Nouns	Verbs
ARK_{IRC}	67.17	78.26	88.26	85.99
$ARK_{Twitter}$	-	-	86.97	88.71
ARK_{Ritter}	68.57	77.29	90.64	85.02
cTAKES	83.93	62.80	93.85	69.08
GENIA	81.56	61.83	92.03	71.01
RASP	-	-	84.59	83.58
Stanford	80.30	73.42	91.89	83.09
SVMTool	69.97	70.04	90.08	80.19
Wapiti	65.64	66.66	87.84	74.87

Table 3: Detailed view of the POS model results focusing on the noun and verb tag groups. The leftmost two columns of figures show accuracies over tags in the respective groups; the rightmost two columns show the accuracies of the same groups if all tags in a group are replaced with a group tag, e.g. *V* for verbs[5].

There are a number of ways we could improve chunking accuracy besides increasing the amount of training data. Although our results do not show a clear trend, Fan et al. (2011) demonstrate that the domain of part-of-speech training data has a significant impact on tagging accuracy, which could potentially impact chunking results if it decreases the number of errors that propagate during chunking. An important problem in that area is dealing with present and past participles, which are almost sure to cause error propagation if mislabelled (as nouns or adjectives, respectively). Participles are more ambiguous in terse contexts lacking auxiliary verbs, which are natural disambiguation indicators. Another direction in processing that could contribute to better chunking is better token and sentence segmentation. Finally, unknown words, which may potentially have the largest impact on chunking accuracy, could be dealt with using a generic solution such as feature expansion based on distributional similarity.

References

S. Abney. 1991. Parsing by chunks. In Robert C. Berwick, Steven P. Abney, and Carol Tenny, editors, *Principle-Based Parsing: Computation and Psycholinguistics*, pages 257–278. Kluwer, Dordrecht.

T. Bentley, C. Price, and P. Brown. 1996. Structural and lexical features of successive versions of the

[5]Note that these results are different from what would be yielded by a classifier trained on data subjected to the same tag substitution.

read codes. In *Proceedings of the Annual Conference of The Primary Health Care Specialist Group of the British Computer Society*, pages 91–103.

T. Briscoe, J. Carroll, and R. Watson. 2006. The second release of the RASP system. In *Proceedings of the COLING/ACL on Interactive Presentation Sessions*, COLING-ACL'06, pages 77–80, Stroudsburg, PA, USA. Association for Computational Linguistics.

N. Chinchor. 1998. Appendix B: Test scores. In *Proceedings of the Seventh Message Understanding Conference (MUC-7)*, Fairfax, VA, April.

J. Clear. 1993. The British National Corpus. In George P. Landow and Paul Delany, editors, *The Digital Word*, pages 163–187. MIT Press, Cambridge, MA, USA.

A. Coden, S. Pakhomov, R. Ando, P. Duffy, and C. Chute. 2005. Domain-specific language models and lexicons for tagging. *Journal of Biomedical Informatics*, 38:422–430.

J.-W. Fan, R. Prasad, R.M. Yabut, R.M. Loomis, D.S. Zisook, J.E. Mattison, and Y. Huang. 2011. Part-of-speech tagging for clinical text: Wall or bridge between institutions? In *American Medical Informatics Association Annual Symposium*, 1, pages 382–391. American Medical Informatics Association.

J.-W. Fan, E. Yang, M. Jiang, R. Prasad, R. Loomis, D. Zisook, J. Denny, H. Xu, and Y. Huang. 2013. Research and applications: Syntactic parsing of clinical text: guideline and corpus development with handling ill-formed sentences. *JAMIA*, 20(6):1168–1177.

J. Giménez and L. Màrquez. 2004. SVMTool: A general POS tagger generator based on support vector machines. In *Proceedings of the 4th LREC*, Lisbon, Portugal.

T. Kudo and Y. Matsumoto. 2001. Chunking with support vector machines. In *Proceedings of the Second Meeting of the North American Chapter of the Association for Computational Linguistics on Language Technologies*, NAACL'01, pages 1–8, Stroudsburg, PA, USA. Association for Computational Linguistics.

T. Kudo and Y. Matsumoto. 2003. Fast methods for kernel-based text analysis. In *Proceedings of the 41st Annual Meeting of the Association for Computational Linguistics*, pages 24–31, Morristown, NJ, USA. Association for Computational Linguistics.

T. Lavergne, O. Cappé, and F. Yvon. 2010. Practical very large scale CRFs. In *Proceedings of the 48th Annual Meeting of the Association for Computational Linguistics*, pages 504–513. Association for Computational Linguistics, July.

M. Marcus, B. Santorini, and M. A. Marcinkiewicz. 1993. Building a large annotated corpus of English: The Penn Treebank. *Computational Linguistics*, 19(2):313–330.

O. Owoputi, B. O'Connor, C. Dyer, K. Gimpel, N. Schneider, and N. Smith. 2013. Improved part-of-speech tagging for online conversational text with word clusters. In *Proceedings of NAACL-HLT*, pages 380–390.

S. Pakhomov, A. Coden, and C. Chute. 2004. Creating a test corpus of clinical notes manually tagged for part-of-speech information. In *Proceedings of the International Joint Workshop on Natural Language Processing in Biomedicine and Its Applications*, JNLPBA'04, pages 62–65, Stroudsburg, PA, USA. Association for Computational Linguistics.

L. Ramshaw and M. Marcus. 1995. Text chunking using transformation-based learning. In *Proceedings of the Third Workshop on Very Large Corpora*, pages 82–94.

E. Sang and S. Buchholz. 2000. Introduction to the CoNLL-2000 shared task: Chunking. In *Proceedings of the 2nd Workshop on Learning Language in Logic and the 4th Conference on Computational Natural Language Learning*, pages 13–14.

G. Savova, J. Masanz, P. Ogren, J. Zheng, S. Sohn, K. Kipper-Schuler, and C. Chute. 2010. Mayo clinical text analysis and knowledge extraction system (cTAKES): architecture, component evaluation and applications. *Journal of the American Medical Informatics Association*, 17(5):507–513.

X. Sun, L.-P. Morency, D. Okanoharay, and J. Tsujii. 2008. Modeling latent-dynamic in shallow parsing: A latent conditional model with improved inference. In *Proceedings of the 22nd International Conference on Computational Linguistics (COLING'08)*, Manchester, UK, August.

K. Toutanova, D. Klein, C. D. Manning, and Y. Singer. 2003. Feature-rich part-of-speech tagging with a cyclic dependency network. In *Proceedings of the 2003 Conference of the North American Chapter of the Association for Computational Linguistics on Human Language Technology - Volume 1*, NAACL'03, pages 173–180, Stroudsburg, PA, USA. Association for Computational Linguistics.

Y. Tsuruoka, Y. Tateishi, J.-D. Kim, T. Ohta, J. McNaught, S. Ananiadou, and J. Tsujii. 2005. Developing a robust part-of-speech tagger for biomedical text. In *Proceedings of the 10th Panhellenic Conference on Advances in Informatics*, PCI'05, pages 382–392, Berlin, Heidelberg. Springer-Verlag.

Decision Style in a Clinical Reasoning Corpus

Limor Hochberg[1] Cecilia O. Alm[1] Esa M. Rantanen[1]
Caroline M. DeLong[1] Anne Haake[2]
1 College of Liberal Arts 2 College of Computing & Information Sciences
Rochester Institute of Technology
lxh6513|coagla|emrgsh|cmdgsh|anne.haake@rit.edu

Abstract

The dual process model (Evans, 2008) posits two types of decision-making, which may be ordered on a continuum from *intuitive* to *analytical* (Hammond, 1981). This work uses a dataset of narrated image-based clinical reasoning, collected from physicians as they diagnosed dermatological cases presented as images. Two annotators with training in cognitive psychology assigned each narrative a rating on a four-point decision scale, from intuitive to analytical. This work discusses the annotation study, and makes contributions for resource creation methodology and analysis in the clinical domain.

1 Introduction

Physicians make numerous diagnoses daily, and consequently clinical decision-making strategies are much discussed (e.g., Norman, 2009; Croskerry, 2003, 2009). Dual process theory proposes that decision-making may be broadly categorized as *intuitive* or *analytical* (Kahneman & Frederick, 2002; Stanovich & West, 2000). Further, scholars argue that decision-making may be ordered on a continuum, with intuitive and analytical at each pole (Hamm, 1988; Hammond, 1981).

Determining the decision strategies used by physicians is of interest because certain styles may be more appropriate for particular tasks (Hammond, 1981), and better suited for expert physicians rather than those in training (Norman, 2009). Language use can provide insight into physician decision style, as linguistic content reflects cognitive processes (Pennebaker & King, 1999).

While most clinical corpora focus on patients or conditions, physician diagnostic narratives have been successfully annotated for conceptual units (e.g., identifying medical morphology or a differential diagnosis), by Womack et al. (2013) and McCoy et al. (2012). Crowley et al. (2013) created an instructional system to detect cognitive biases in clinical decision-making, while Coderre et al. (2003) used protocol analysis on think-aloud diagnostic narratives, and found that features of intuitive reasoning implied diagnostic accuracy.

In this study, speech data were collected from physicians as they diagnosed dermatological cases presented to them as images. Physician verbalizations were annotated for decision style on a four-point scale from intuitive to analytical (Figure 1). Importantly, cognitive psychologists were brought into the loop for decision style annotation, to take advantage of their expertise in decision theory.

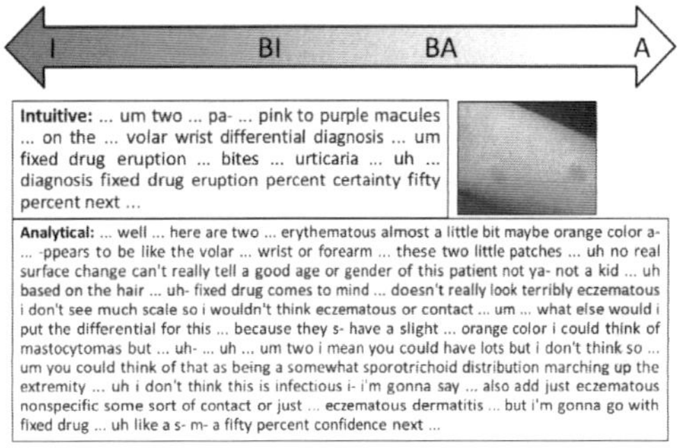

Figure 1: The decision-making continuum, showing the four-point rating scale. The example narratives were by two physicians for the same image (used with permission from Logical Images, Inc.), both correct in diagnosis. *(I=Intuitive, BI=Both-Intuitive, BA=Both-Analytical, A=Analytical).*

This work describes a thorough methodology applied in annotating a corpus of diagnostic narratives for decision style. The corpus is a unique resource – the first of its kind – for studying and modeling clinical decision style or for developing instructional systems for training clinicians to assess their reasoning processes.

This study attempts to capture empirically decision-making constructs that are much-

Proceedings of the 2014 Workshop on Biomedical Natural Language Processing (BioNLP 2014), pages 83–87,
Baltimore, Maryland USA, June 26-27 2014.

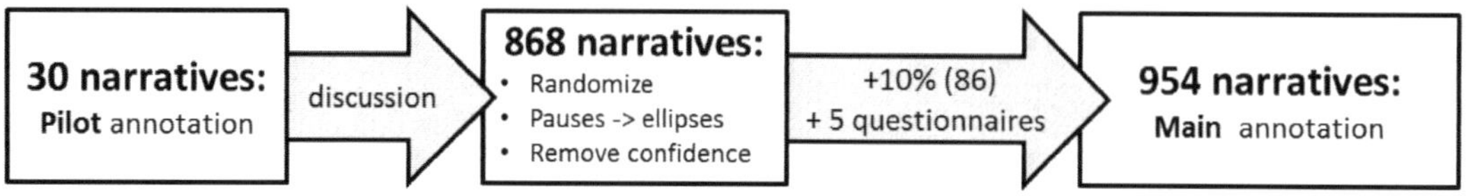

Figure 2: **Overview of annotation methodology.** Conclusions from the pilot study enhanced the main annotation study. To ensure high-quality annotation, narratives appeared in random order, and 10% (86) of narratives were duplicated and evenly distributed in the annotation data, to later assess intra-annotator reliability. Questionnaires were also interspersed at 5 equal intervals to study annotator strategy.

discussed theoretically. Thus, it responds to the need for investigating subjective natural language phenomena (Alm, 2011). The annotated corpus is a springboard for decision research in medicine, as well as other mission-critical domains in which good decisions save lives, time, and money.

Subjective computational modeling is particularly challenging because often, no real 'ground truth' is available. Decision style is such a *fuzzy* concept, lacking clear boundaries (Hampton, 1998), and its recognition develops in psychologists over time, via exposure to knowledge and practice in cognitive psychology. Interpreting fuzzy decision categories also depends on mental models which lack strong intersubjective agreement. This is the nature, and challenge, of capturing understandings that emerge organically.

This work's contributions include (1) presenting a distinct clinical resource, (2) introducing a robust method for fuzzy clinical annotation tasks, (3) analyzing the annotated data comprehensively, and (4) devising a new metric that links annotated behavior to clinicians' decision-making profiles.

2 Corpus Description

In an experimental data-collection setting, 29 physicians (18 residents, 11 attendings) narrated their diagnostic thought process while inspecting 30 clinical images of dermatological cases, for a total of 868[1] narratives. Physicians described observations, differential and final diagnoses, and confidence (out of 100%) in their final diagnosis. Later, narratives were assessed for correctness (based on final diagnoses), and image cases were evaluated for difficulty by a dermatologist.

3 Corpus Annotation of Decision Style

The corpus was annotated for decision style in a pilot study and then a main annotation study (Figure 2).[2] Two annotators with graduate training in cognitive psychology independently rated each narrative on a four-point scale from *intuitive* to *analytical* (Figure 1). The two middle labels reflect the presence of both styles, with intuitive (*BI*) or analytical (*BA*) reasoning being more prominent. Since analytical reasoning involves detailed examination of alternatives, annotators were asked to avoid using length as a proxy for decision style.

After the pilot, the annotators jointly discussed disagreements with one researcher. Inter-annotator reliability, measured by linear weighted kappa (Cohen, 1968), was 0.4 before and 0.8 after resolution; the latter score may be an upper bound on agreement for clinical decision-making annotation. As both annotators reported using physician-provided confidence to judge decision style, in subsequent annotation confidence mentions had been removed if they appeared after the final diagnosis (most narratives), or, if intermixed with diagnostic reasoning, replaced with dashes. Finally, silent pauses[3] were coded as ellipses to aid in the human parsing of the narratives.

4 Quantative Annotation Analysis

Table 1 shows the annotator rating distributions.[4]

	I	BI	BA	A
A1	89	314	340	124
A2	149	329	262	127

Table 1: The distribution of ratings across the 4-point decision scale. *I=Intuitive, BI=Both-Intuitive, BA=Both-Analytical, A=Analytical; A1=Annotator 1, A2=Annotator 2; N=867.*

Though Annotator 1's ratings skew slightly more analytical than Annotator 2, a Kolmogorov-

[1] Two physicians skipped 1 image during data collection.

[2] Within a reasonable time frame, the annotations will be made publicly available as part of a corpus release.

[3] Above around 0.3 seconds (see Lövgren & Doorn, 2005).

[4] N = 867 after excluding a narrative that, during annotation, was deemed too brief for decision style labeling.

Factor	A1 (Avg)	A1 (SD)	A2 (Avg)	A2 (SD)
Switching between decision styles	1.0	0.0	3.6	0.9
Timing of switch between decision styles	1.6	0.5	4.2	0.4
Silent pauses (...)	2.0	0.0	3.6	0.5
Filled pauses (e.g. *uh, um*)	2.0	0.7	3.6	0.5
Rel. (similarity) of final & differential diagnosis	2.8	0.4	3.2	0.8
Use of logical rules and inference	3.2	0.8	2.2	0.4
False starts (in speech)	3.4	0.9	2.4	0.9
Automatic vs. controlled processing	3.4	0.5	4.0	0.0
Holistic vs. sequential processing	3.6	0.5	4.4	0.5
No. of diagnoses in differential diagnoses	4.0	0.0	1.6	0.5
Word choice	4.0	0.7	2.6	0.5
Rel. (similarity) of final & first-mentioned diagnosis	4.0	0.0	4.0	0.0
Perceived attitude	4.0	0.7	4.0	0.0
Rel. timing of differential diagnosis in the narrative	4.2	0.8	2.8	0.8
Degree of associative (vs. linear, ordered) processing	4.2	0.4	3.8	0.4
Use of justification (e.g. *X because Y*)	4.2	0.4	4.0	0.0
Perceived confidence	4.4	0.5	4.2	0.4

Table 3: Annotators rated each of the listed factors as to how often they were used in annotation, on a 5-point Likert scale from *for no narratives* (1) to *for all narratives (5)*. (Some factors slightly reworded.)

Smirnov test showed no significant difference between the two distributions ($p = 0.77$).

	WK	%FA	%FA+ 1	*N*
A1 - A2	.43	50%	94%	867
A1 - A1	.64	67%	100%	86
A2 - A2	.43	50%	95%	86

Table 2: Inter- and intra-annotator reliability, measured by linear weighted kappa (WK), percent full agreement (%FA); and full plus within 1-point agreement (%FA+1). Intra-annotator reliability was calculated for the narratives rated twice, and inter-annotator reliability on the initial ratings.

As shown in Table 2, reliability was moderate to good (Altman, 1991), and inter-annotator agreement was well above chance (25%). Indeed, annotators were in full agreement, or agreed within one rating on the continuum, on over 90% of narratives. This pattern reveals fuzzy category boundaries but sufficient regularity so as to be measurable. This is in line with subjective natural language phenomena, and may be a consequence of imposing discrete categories on a continuum.[5] Annotator 1 had better intra-annotator reliability, perhaps due to differences in annotation strategy.

[5]Nonetheless, affect research has shown that scalar representations are not immune to variation issues (Alm, 2009).

5 Annotator Strategy Analysis

Five questionnaires evenly spaced among the narratives asked annotators to rate how often they used various factors in judging decision style (Table 3). Factors were chosen based on discussion with the annotators after the pilot, and referred to in descriptions of decision styles in the annotator instructions; the descriptions were based on characteristics of each style in the cognitive psychology literature (e.g., Evans, 2008). Factors with high variability (SD columns in Table 3) reveal changes in annotator strategy over time, and factors that may influence intra-annotator reliability.

Both annotators reported using the *rel. (similarity) of final & first-mentioned diagnosis*, as well as *perceived attitude*, *perceived confidence*, and *use of justification*, to rate most narratives. Types of *processing* were used by both sometimes; this is important since these are central to the definitions of decision style in decision-making theory.

Differences in strategies allow for the assessment of annotators' individual preferences. Annotator 1 often considered the *no. of diagnoses in the differential*, and *rel. timing of the differential*, but Annotator 2 rarely attended to them; the opposite pattern occurred with respect to *switching between decision styles*, and the *timing of the switch*.

The shared high factors reveal those consistently linked to interpreting decision style, despite the concept's fuzzy boundaries. In contrast, the idiosyncratic high factors reveal starting points for understanding fuzzy perception, and for further calibrating inter-annotator reliability.

6 Narrative Case Study

Examining particular narratives is also instructive. Of the 86 duplicated narratives with two ratings per annotator, *extreme agreement* occurred for 22 cases (26%), meaning that all four ratings were exactly the same.[6] Figure 3 (top) shows such a case of intuitive reasoning: a quick decision without reflection or discussion of the differential. Figure 3 (middle) shows a case of analytical reasoning: consideration of alternatives and logical inference.

Agr (I): ... there's a ... brown papule with telangiectasias on the ... nasal tip ... uh the differential includes a pigmented basal cell melanoma ... nevus ... and the diagnosis is melanoma **(diagnosis incorrect)**
Agr (A): ... okay so a large ... purple ... uh ... mass ... on a face ... no ... it's on the foot ... or the ... yeah ... um ... yeah it would **depend** a lot on how well it blanches ... you want- wanna ... feel that ... um ... could be just a ... hemangioma ... could be a ... basal cell skin cancer could be a melanoma ... um uh might be one of those things you wanna ... toughen the uh ... the edge of it ... has a bit of a ... pearly look to it but i don't know if that's just again from ... being on a foot ... and uh ... and having more uh ... hydrostatic pressures there ... um -**cause** it's mostly ... uh purple ... it could be a you know angiosarcoma um but it's a little on the small side ... um ... you know common things being common go with the uh ... hemangioma ... as the number one thought ... with uh ... m- basal cell skin cancer being the second hemangioma again **(diagnosis incorrect)**
DisAgr (A, I): ... uh uh ... think we're on a foot you see some scale at the bottom makes me think there's little fungus there but ... looks like the thing that they took the picture of is a purple irregular tumor ... um ... has very ill-distinct borders with surrounding red areas ... **it's so purple it makes me think of a vascular tumor ... so i think kaposi's sarcoma is most likely** ... could be a melanoma ... could be a metastatic renal cell tumor ... my best guess is that this is kaposi's sarcoma **(diagnosis incorrect)**

Figure 3: Narratives for which annotators were in *full agreement* on *I* (top) and *A* (middle) ratings, vs. in *extreme disagreement* (bottom).

In the full data set (initial ratings), there were 50 cases (6%) of 2-point inter-annotator disagreement and one case of 3-point inter-annotator disagreement (Figure 3, bottom). This latter narrative was produced by an attending (experienced physician), 40% confident and incorrect in the final diagnosis. Annotator 1 rated it analytical, while Annotator 2 rated it intuitive. This is in line with Annotator 1's preference for analytical ratings (Table 1). Annotator 1 may have viewed this pattern of *observation → conclusion* as logical reasoning, characteristic of analytical reasoning. Annotator 2 may instead have interpreted the phrase *it's so purple it makes me think of a vascular tumor...so i think [...]* as intuitive, due to the *makes me think* comment, indicating associative reasoning, characteristic of intuitive thinking. This inter-annotator contrast may reflect Annotator 1's greater reported use of the factor *logical rules and inference* (Table 3).

[6]There were no cases where all four labels differed, further emphasizing the phenomenon's underlying regularity.

7 Physician Profiles of Decision Style

Annotations were also used to characterize physicians' preferred decision style. A decision score was calculated for each physician as follows:

$$d_p = \frac{1}{2n}\sum_{i=1}^{n}(r_{A1_i} + r_{A2_i}) \qquad (1)$$

where p is a physician, r is a rating, n is total images, and $A1, A2$ the annotators. Annotators' initial ratings were summed – from 1 for *Intuitive* to 4 for *Analytical* – for all image cases for each physician, and divided by 2 times the number of images, to normalize the score to a 4-point scale. Figure 4 shows the distribution of decision scores across *residents* and experienced *attendings*.

Residents exhibit greater variability in decision style. While this might reflect that residents were the majority group, it suggests that differences in expertise are linked to decision styles; such differences hint at the potential benefits that could come from preparing clinical trainees to self-monitor their use of decision style. Interestingly, the overall distribution is skewed, with a slight preference for analytical decision-making, and especially so for attendings. This deserves future attention.

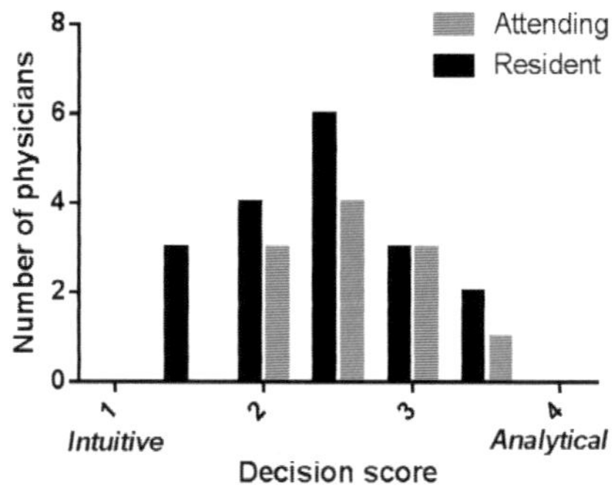

Figure 4: Decision score distribution by expertise.

8 Conclusion

This study exploited two layers of expertise: physicians produced diagnostic narratives, and trained cognitive psychologists annotated for decision style. This work also highlights the importance of understanding annotator strategy, and factors influencing annotation, when fuzzy categories are involved. Future work will examine the links between decision style, expertise, and diagnostic accuracy or difficulty.

Acknowledgements

Work supported by a CLA Faculty Dev. grant, Xerox award, and NIH award R21 LM01002901. Many thanks to annotators and reviewers.

This content is solely the responsibility of the authors and does not necessarily represent the official views of the National Institutes of Health.

References

Alm, C. O. (2009). *Affect in text and speech.* Saarbrücken: VDM Verlag.

Alm, C. O. (2011, June). Subjective natural language problems: Motivations, applications, characterizations, and implications. In *Proceedings of the 49th Annual Meeting of the Association for Computational Linguistics: Human Language Technologies: short papers-Volume 2* (pp. 107-112). Association for Computational Linguistics.

Altman, D. (1991). *Practical statistics for medical research.* London: Chapman and Hall.

Coderre, S., Mandin, H., Harasym, P. H., & Fick, G. H. (2003). Diagnostic reasoning strategies and diagnostic success. *Medical Education, 37*(8), 695-703.

Cohen, J. (1968). Weighted kappa: Nominal scale agreement provision for scaled disagreement or partial credit. *Psychological Bulletin, 70*(4), 213-220.

Crowley, R. S., Legowski, E., Medvedeva, O., Reitmeyer, K., Tseytlin, E., Castine, M., ... & Mello-Thoms, C. (2013). Automated detection of heuristics and biases among pathologists in a computer-based system. *Advances in Health Sciences Education, 18*(3), 343-363.

Croskerry, P. (2003). The importance of cognitive errors in diagnosis and strategies to minimize them. *Academic Medicine, 78*(8), 775-780.

Croskerry, P. (2009). A universal model of diagnostic reasoning. *Academic Medicine, 84*(8), 1022-1028.

Evans, J. (2008). Dual-processing accounts of reasoning, judgement and social cognition. *Annual Review of Psychology, 59*, 255-278.

Hamm, R. M. (1988). Clinical intuition and clinical analysis: Expertise and the cognitive continuum. In J. Dowie & A.S. Elstein (Eds.), *Professional judgment: A reader in clinical decision making* (pp. 78-105). Cambridge, England: Cambridge University Press.

Hammond, K. R. (1981). *Principles of organization in intuitive and analytical cognition (Report #231).* Boulder, CO: University of Colorado, Center for Research on Judgment & Policy.

Hampton, J. A. (1998). Similarity-based categorization and fuzziness of natural categories. *Cognition, 65*(2), 137-165.

Kahneman, D., & Frederick, S. (2002). Representativeness revisited: Attribute substitution in intuitive judgment. In T. Gilovich, D. Griffin, & D. Kahneman (Eds.), *Heuristics of intuitive judgment: Extensions and applications* (pp. 49-81). New York, NY: Cambridge University Press.

Lövgren, T., & Doorn, J. V. (2005). Influence of manipulation of short silent pause duration on speech fluency. In *Proceedings of Disfluency in Spontaneous Speech Workshop* (pp. 123-126). International Speech Communication Association.

McCoy, W., Alm, C. O., Calvelli, C., Li, R., Pelz, J. B., Shi, P., & Haake, A. (2012, July). Annotation schemes to encode domain knowledge in medical narratives. In *Proceedings of the 6th Linguistic Annotation Workshop* (pp. 95-103). Association for Computational Linguistics.

Norman, G. (2009). Dual processing and diagnostic errors. *Advances in Health Sciences Education, 14*(1), 37-49.

Pennebaker, J. W., & King, L. A. (1999). Linguistic styles: Language use as an individual difference. *Journal of Personality and Social Psychology, 77*(6), 1296-1312.

Stanovich, K. E., & West, R. F. (2000). Individual differences in reasoning: Implications for the rationality debate? *Behavioral and Brain Sciences, 23*, 645-665.

Womack, K., Alm, C. O., Calvelli, C., Pelz, J. B., Shi, P., and Haake, A. (2013, August). Using linguistic analysis to characterize conceptual units of thought in spoken medical narratives. In *Proceedings of Interspeech 2013* (pp. 3722-3726). International Speech Communication Association.

Temporal Expressions in Swedish Medical Text – A Pilot Study

Sumithra Velupillai
Department of Computer and Systems Sciences
Stockholm University
Sweden
sumithra@dsv.su.se

Abstract

One of the most important features of health care is to be able to follow a patient's progress over time and identify events in a temporal order. We describe initial steps in creating resources for automatic temporal reasoning of Swedish medical text. As a first step, we focus on the identification of temporal expressions by exploiting existing resources and systems available for English. We adapt the HeidelTime system and manually evaluate its performance on a small subset of Swedish intensive care unit documents. On this subset, the adapted version of HeidelTime achieves a precision of 92% and a recall of 66%. We also extract the most frequent temporal expressions from a separate, larger subset, and note that most expressions concern parts of days or specific times. We intend to further develop resources for temporal reasoning of Swedish medical text by creating a gold standard corpus also annotated with events and temporal links, in addition to temporal expressions and their normalised values.

1 Introduction

One of the most important features of health care is to be able to follow patient progress over time and identify clinically relevant events in a temporal order. In medical records, temporal information is stored with explicit timestamps, but it is also documented in free text in the clinical narratives. To meet our overall goal of building accurate and useful information extraction systems in the health care domain, our aim is to build resources for temporal reasoning in Swedish clinical text. For instance, in the example sentence ***MR-undersökningen av skallen*** *igår visade att den vä-sidiga förändringen i thalamus minskat i volym.* ("The **MRI-scan of the scull** yesterday showed that the left (abbreviated) side change in thalamus has decreased in volume"), a temporal reasoning system should extract the **event** (**MRI-scan of the scull**) and the temporal expression (yesterday), and be able to normalise the time expression to a specific date and classify the temporal relation.

In this pilot study we focus on the identification of temporal expressions, utilising existing resources and systems available for English. A temporal expression is defined as any mention of dates, times, durations, and frequencies, e.g. "April 2nd", "10:50am", "five hours ago", and "every 2 hours". When successfully identifying such expressions, subsequent anchoring in time is made possible.

Although English and Swedish are both Germanic languages, there are some differences that are important to take into account when adapting existing solutions developed for English to Swedish, e.g. Swedish is more inflective and is more compounding than English.

The purpose of this study is to initiate our work on temporal reasoning for Swedish, and to evaluate existing solutions adapted to Swedish. These are our first steps towards the creation of a reference standard that can be used for evaluation of future systems.

2 Background

Temporal reasoning has been the focus of several international natural language processing (NLP) challenges in the general domain such as ACE[1], TempEval-2 and 3 (Verhagen et al., 2010; UzZaman et al., 2013), and in the clinical domain through the 2012 i2b2 challenge (Sun et al., 2013). Most previous work has been performed on En-

[1] http://www.itl.nist.gov/iad/mig/tests/ace/

Proceedings of the 2014 Workshop on Biomedical Natural Language Processing (BioNLP 2014), pages 88–92,
Baltimore, Maryland USA, June 26-27 2014.

glish documents, but the TempEval series have also included other languages, e.g. Spanish. For temporal modelling, the TimeML (Pustejovsky et al., 2010) guidelines are widely used. The TimeML standard denotes events (EVENT), temporal expressions (TIMEX3) and temporal relations (TLINK).

For English, several systems have been developed for all or some of these subtasks, such as the TARSQI Toolkit (Verhagen et al., 2005) and SUTime (Chang and Manning, 2012). Both these tools are rule-based, and rely on regular expressions and gazetteers. The TARSQI Toolkit has also been developed for the clinical domain: MedTTK (Reeves et al., 2013).

In other domains, and for other languages, HeidelTime (Strötgen and Gertz, 2012) and TIMEN (Llorens et al., 2012) are examples of other rule-based systems. These are also developed to be easily extendable to new domains and languages. HeidelTime ranked first in the TempEval-3 challenge on TIMEX3:s, resulting in an $F1$ of 77.61 for the task of correctly identifying and normalising temporal expressions.

HeidelTime was also used in several participating systems in the i2b2 challenge (Lin et al., 2013; Tang et al., 2013; Grouin et al., 2013) with success. Top results for correctly identifying and normalising temporal expressions in the clinical domain are around 66 $F1$ (Sun et al., 2013). The system has also been adapted for French clinical text (Hamon and Grabar, 2014).

3 Methods

The HeidelTime system was chosen for the initial development of a Swedish temporal expression identifier. Given that its architecture is designed to be easily extendible for other languages as well as domains, and after reviewing alternative existing systems, we concluded that it was suitable for this pilot study.

3.1 Data

We used medical records from an intensive care unit (ICU) from the Stockholm EPR Corpus, a clinical database from the Stockholm region in Sweden[2] (Dalianis et al., 2012). Each medical record (document) contains all entries (notes) about one patient a given day. The document contains notes written by both physicians and nurses. They also contain headings (e.g. *Daganteckning* ("Daily note"), *Andning* ("Breathing")) and timestamps for when a specific note/heading was recorded in the medical record system. These are excluded in this analysis.

Three subsets from this ICU dataset were used: 1) two randomly selected documents were used for analysing and identifying domain specific time expressions and regular expressions to be added in the adaptation of HeidelTime (development set), 2) a random sample of ten documents was used for manual analysis and evaluation (test set), and 3) a set of 100 documents was also extracted for the purpose of empirically studying the types of temporal expressions found in the data by the adapted system (validation set).

3.2 Adaptation of HeidelTime and Evaluation

The available resources (keywords and regular expression rules) in the HeidelTime system were initially translated automatically (Google translate[3]) and manually corrected. Regular expressions were modified to handle Swedish inflections and other specific traits. An initial analysis on two separate, randomly selected ICU notes (development set) was performed, as a first step in adapting for both the Swedish language and the clinical domain.

Results on the system performance were manually evaluated on the test set by one computational linguistics researcher by analysing system outputs: adding annotations when the system failed to identify a temporal expression, and correcting system output errors. A contingency table was created for calculating precision, recall and $F1$, the main outcome measures. Moreover, the top most frequent temporal expressions found by the system on a separate set were extracted (validation set), for illustration and analysis purposes.

4 Results

We report general statistics for the ICU corpus, results from the adaptation and evaluation of HeidelTime for Swedish (HTSwe) on the test set, and the most frequent temporal expressions found by HTSwe in a separate set of 100 ICU documents (validation set).

[2] Study approved by the Regional Ethical Review Board in Stockholm (Etikprövningsnämnden i Stockholm), permission number 2012/834-31/5

[3] http://translate.google.se

4.1 Data: ICU corpus

General statistics for the test set used in this study is shown in Table 1. On average, each document consists of 54.6 sentences, and each sentence contains on average 8.7 tokens (including punctuation). We observe that some sentences are very short (min = 1), and there is great variability in length, as can be seen through the standard deviation.

	#	min - max	avg ± std
Sentences /document	540	35 - 80	54.6±14.1
Tokens /sentence	4749	1 - 52	8.7±5.7

Table 1: General statistics for the test set (ten ICU documents) used in this study. Minimum, maximum, average and standard deviation for sentences per document and tokens (including punctuation) per sentence.

4.2 Adaptation and evaluation of HeidelTime: HTSwe

The main modifications required in the adaptation of HeidelTime to Swedish (HTSwe) involved handling definite articles and plurals, e.g. adding *eftermiddag(en)?(ar)?(na)?* ("afternoon", "the afternoon"/"afternoons"/"the afternoons"). From the analysis of the small development set, some abbreviations were also added, e.g. *em* ("afternoon"). Regular expressions for handling typical ways dates are written in Swedish were added, e.g. "020812" and "31/12 -99" (day, month, year). In order to avoid false positives, a rule for handling measurements that could be interpreted as years (e.g. *1900 ml*) was also added (a negative rule).

Results from running HTSwe on the test set are shown in Table 2. HTSwe correctly identified 105 temporal expressions, but missed 55 expressions that should have been marked, and classified 9 expressions erroneously. In total, there are 160 TIMEX3s. Overall performance was 92% precision, 65% recall and $F1 = 77\%$.

The main errors were due to faulty regular expressions for times, e.g. *13-tiden* ("around 13 PM) and missing keywords such as *dygn* ("day" - a word to indicate a full day, i.e. 24 hours) and *lunchtid* ("around lunch"). Some missing keywords were specific for the clinical domain, e.g. *efternatten* ("the after/late night", typical for shift indication). There were also some partial errors. For instance, *i dag* ("today") was only included with the spelling *idag* in the system, thus generating a TIMEX3 output only for *dag*.

	TIMEX3 Annotator	**Other** Annotator	$\sum$
TIMEX3 HTSwe	105	9	114
Other HTSwe	55	4580	4635
$\sum$	160	4589	4749

Table 2: Contingency table, TIMEX3 annotations by the annotator and the adapted HeidelTime system for Swedish (HTSwe) on the test set. "Other" means all other tokens in the corpus. These results yield a precision of 92%, a recall of 66%, and $F1$ = 77% for HTSwe.

On the validation set, 168 unique time expressions were found by the system, and 1,178 in total. The most frequent expressions all denote parts of days, e.g. *idag* ("today"), *nu* ("now"), and *natten* ("the night"), see Table 3. Specific times (mostly specific hours) were also very common. Thus, there were many translated expressions in the HeidelTime system that never occurred in the data.

TIMEX3	N	%
idag ("today")	164	14%
nu ("now")	132	11%
natten ("the night")	117	10%
morgonen ("the morning")	96	8%
em ("afternoon", abbreviated)	82	7%
kvällen ("the evening")	74	6%
igår ("yesterday")	49	4%
fm ("morning", abbreviated)	34	3%
morgon ("morning")	30	3%
natt ('night")	26	2%
Total	1178	100%

Table 3: Most frequent (top ten, descending order) TIMEX3s found by HTSwe on the validation set (100 ICU documents). Total = all TIMEX3:s found by HTSwe in the entire validation set. There were 168 unique TIMEX3s in the validation set.

5 Discussion and Conclusion

We perform an initial study on automatic identification of temporal expressions in Swedish clinical text by translating and adapting the HeidelTime system, and evaluating performance on Swedish ICU records. Results show that precision is high (92%), which is promising for our future development of a temporal reasoning system for Swedish. The main errors involve regular expressions for time and some missing keywords; these expressions will be added in our next iteration in this work. Our results, $F1 = 77\%$, are lower than state-of-the-art systems for English clinical text, where the top-performing system in the 2010 i2b2 Challenge achieved 90% $F1$ for TIMEX3 spans (Sun et al., 2013). However, given the small size of this study, results are encouraging, and we have created a baseline system which can be used for further improvements.

The adaptation and translation of HeidelTime involved extending regular expressions and rules to handle Swedish inflections and specific ways of writing dates and times. Through a small, initial analysis on a development set, some further additions and modifications were made, which led to the correct identification of common TIMEX3s present in this type of document. A majority of the expressions translated from the original system was not found in the data. Hence, it is worthwhile analysing a small subset to inform the adaptation of HeidelTime.

The ICU notes are an interesting and suitable type of documentation for temporal reasoning studies, as they contain notes on the progress of patients in constant care. However, from the results it is evident that the types of TIMEX3 expressions are rather limited and mostly refer to parts of days or specific times. Moreover, as recall was lower (66%), there is clearly room for improvement. We plan to extend our study to also include other report types.

5.1 Limitations

There are several limitations in this study. The corpus is very small, and evaluated only by one annotator, which limits the conclusions that can be drawn from the analysis. For the creation of a reference standard, we plan to involve at least one clinician, in order to get validation from a domain expert, and to be able to calculate inter-annotator agreement. The size of the corpus will also be increased. We have not evaluated performance on TIMEX3 normalisation, which, of course, is crucial for an accurate temporal reasoning system. For instance, we have not considered the category *Frequency*, which is essential in the clinical domain to capture e.g. medication instructions and dosages. Moreover, we have not annotated and evaluated *events*. This is perhaps the most important part of a temporal reasoning system. We plan to utilise existing named entity taggers developed in our group as a pre-annotation step in the creation of our reference standard. The last step involves annotating temporal links (TLINK) between events and TIMEX3:s. We believe that part-of-speech (PoS) and/or syntactic information will be a very important component in an end-to-end system for this task. We plan to tailor an existing Swedish PoS tagger, to better handle Swedish clinical text.

5.2 Conclusion

Our main finding is that it is feasible to adapt HeidelTime to the Swedish clinical domain. Moreover, we have shown that the parts of days and specific times are the most frequent temporal expressions in Swedish ICU documents.

This is the first step towards building resources for temporal reasoning in Swedish. We believe these results are useful for our continued endeavour in this area. Our next step is to add further keywords and regular expressions to improve recall, and to evaluate TIMEX3 normalisation. Following that, we will annotate events and temporal links.

To our knowledge, this is the first study on temporal expression identification in Swedish clinical text. All resulting gazetteers and guidelines in our future work on temporal reasoning in Swedish will be made publicly available.

Acknowledgments

The author wishes to thank the anonymous reviewers for invaluable comments on this manuscript. Thanks also to Danielle Mowery and Dr. Wendy Chapman for all their support. This work was partially funded by Swedish Research Council (350-2012-6658) and Swedish Fulbright Commission.

References

Angel X. Chang and Christopher Manning. 2012. SUTime: A library for recognizing and normalizing time expressions. In Nicoletta Calzolari (Conference Chair), Khalid Choukri, Thierry Declerck,

Mehmet Uur Doan, Bente Maegaard, Joseph Mariani, Jan Odijk, and Stelios Piperidis, editors, *Proceedings of the Eight International Conference on Language Resources and Evaluation (LREC'12)*, Istanbul, Turkey, may. European Language Resources Association (ELRA).

Hercules Dalianis, Martin Hassel, Aron Henriksson, and Maria Skeppstedt. 2012. Stockholm EPR Corpus: A Clinical Database Used to Improve Health Care. In Pierre Nugues, editor, *Proc. 4th SLTC, 2012*, pages 17–18, Lund, October 25-26.

Cyril Grouin, Natalia Grabar, Thierry Hamon, Sophie Rosset, Xavier Tannier, and Pierre Zweigenbaum. 2013. Eventual situations for timeline extraction from clinical reports. *JAMIA*, 20:820–827.

Thierry Hamon and Natalia Grabar. 2014. Tuning HeidelTime for identifying time expressions in clinical texts in English and French. In *Proceedings of the 5th International Workshop on Health Text Mining and Information Analysis (Louhi)*, pages 101–105, Gothenburg, Sweden, April. Association for Computational Linguistics.

Yu-Kai Lin, Hsinchun Chen, and Randall A. Brown. 2013. MedTime: A temporal information extraction system for clinical narratives. *Journal of Biomedical Informatics*, 46:20–28.

Hector Llorens, Leon Derczynski, Robert Gaizauskas, and Estela Saquete. 2012. TIMEN: An Open Temporal Expression Normalisation Resource. In Nicoletta Calzolari (Conference Chair), Khalid Choukri, Thierry Declerck, Mehmet Uur Doan, Bente Maegaard, Joseph Mariani, Jan Odijk, and Stelios Piperidis, editors, *Proceedings of the Eight International Conference on Language Resources and Evaluation (LREC'12)*, Istanbul, Turkey, may. European Language Resources Association (ELRA).

James Pustejovsky, Kiyong Lee, Harry Bunt, and Laurent Romary. 2010. ISO-TimeML: An International Standard for Semantic Annotation. In *Proceedings of the Seventh International Conference on Language Resources and Evaluation (LREC'10)*, Valletta, Malta, may. European Language Resources Association (ELRA).

Ruth M. Reeves, Ferdo R. Ong, Michael E. Matheny, Joshua C. Denny, Dominik Aronsky, Glenn T. Gobbel, Diane Montella, Theodore Speroff, and Steven H. Brown. 2013. Detecting temporal expressions in medical narratives. *International Journal of Medical Informatics*, 82:118–127.

Jannik Strötgen and Michael Gertz. 2012. Temporal Tagging on Different Domains: Challenges, Strategies, and Gold Standards. In *Proceedings of the Eigth International Conference on Language Resources and Evaluation (LREC'12)*, pages 3746–3753. ELRA.

Weiyi Sun, Anna Rumshisky, and Özlem Uzuner. 2013. Evaluating temporal relations in clinical text: 2012 i2b2 Challenge. *JAMIA*, 20(5):806–813.

Buzhou Tang, Yonghui Wu, Min Jiang, Yukun Chen, Joshua C Denny, and Hua Xu. 2013. A hybrid system for temporal information extraction from clinical text. *JAMIA*, 20:828–835.

Naushad UzZaman, Hector Llorens, Leon Derczynski, James Allen, Marc Verhagen, and James Pustejovsky. 2013. SemEval-2013 Task 1: TempEval-3: Evaluating Time Expressions, Events, and Temporal Relations. In *Second Joint Conference on Lexical and Computational Semantics (*SEM), Volume 2: Proceedings of the Seventh International Workshop on Semantic Evaluation (SemEval 2013)*, pages 1–9, Atlanta, Georgia, USA, June. Association for Computational Linguistics.

Marc Verhagen, Inderjeet Mani, Roser Sauri, Robert Knippen, Seok Bae Jang, Jessica Littman, Anna Rumshisky, John Phillips, and James Pustejovsky. 2005. Automating Temporal Annotation with TARSQI. In *Proceedings of the ACL 2005 on Interactive Poster and Demonstration Sessions*, ACLdemo '05, pages 81–84, Stroudsburg, PA, USA. Association for Computational Linguistics.

Marc Verhagen, Roser Saurí, Tommaso Caselli, and James Pustejovsky. 2010. SemEval-2010 Task 13: TempEval-2. In *Proceedings of the 5th International Workshop on Semantic Evaluation*, SemEval '10, pages 57–62, Stroudsburg, PA, USA. Association for Computational Linguistics.

A repository of semantic types in the MIMIC II database clinical notes

Richard M. Osborne
Computational Bioscience
University of Colorado
School of Medicine
richard.osborne@ucdenver.edu

Alan R. Aronson
National Library of Medicine
Bethesda, MD
alan@nlm.nih.gov

K. Bretonnel Cohen
Computational Bioscience
University of Colorado
School of Medicine
kevin.cohen@gmail.com

Abstract

The MIMIC II database contains 1,237,686 clinical documents of various kinds. A common task for researchers working with this database is to run MetaMap, which uses the UMLS Metathesaurus, on those documents to identify specific semantic types of entities mentioned in them. However, this task is computationally expensive and time-consuming. Research in many groups could be accelerated if there were a community-accessible set of outputs from running MetaMap on this document collection, cached and available on the MIMIC-II website. This paper describes a repository of all MetaMap output from the MIMIC II database, publicly available, assuming compliance with usage agreements required by UMLS and MIMIC-II. Additionally, software for manipulating MetaMap output, available on SourceForge with a liberal Open Source license, is described.

1 Introduction

1.1 The MIMIC II database and its textual contents

The Multiparameter Intelligent Monitoring in Intensive Care II (MIMIC-II) database is a public-access intensive care unit database that contains a broad array of information for over 33,000 patients. The data were collected over a 7 year period, beginning in 2001 from Boston's Beth Israel Deaconess Medical Center (Saeed et al, 2011; Goldberger et al., 2000).

Of particular interest are the 1,237,686 clinical documents, which are broadly classified into the following four groups: MD notes, discharge summaries, radiology reports and nursing/other. Each free-text note contains information describing such things as a given patient's health, illnesses, treatments and medications, among others.

1.2 Motivation for the resource: MetaMap runtimes

Part of the motivation for making this resource publicly available is that considerable resources must be expended to process it; if multiple groups can share the output of one processing run, the savings across the community as a whole could be quite large. To illustrate why this would be valuable from a resources perspective, we provide here some statistics on the performance of MetaMap.

Random samples of each category (10% each) were chosen and Monte Carlo simulation was performed (1,000 iterations per note) to obtain the running times presented below. The clinical notes ranged from a minimum of 0 words to a maximum of 6,684 (some of the notes were 0 bytes because the note for a particular patient and day contained no text). The mean, median and mode per document processed by MetaMap were 17, 5 and 2 seconds, respectively, with a minimum of 1 and a maximum of 216 seconds.

Figure 1 below plots the number of words against processing times in seconds for each of the of notes, sampled as mentioned above.

The majority of the processing was done on a Sun Fire X4600M2 server with 16 (4 x Quad-Core AMD Opteron(tm) Processor 8356 cores, 2.3GHz), 128GB memory and 12 TB of disk storage, currently running Fedora Core 17 Linux. (An Apple MacBook Pro and a Windows desktop server were also used to speed processing. The analysis of the random sample of notes was performed in its entirety on the Sun machine, thereby providing consistent results for the data in Figure 1.)

Proceedings of the 2014 Workshop on Biomedical Natural Language Processing (BioNLP 2014), pages 93–97,
Baltimore, Maryland USA, June 26-27 2014.

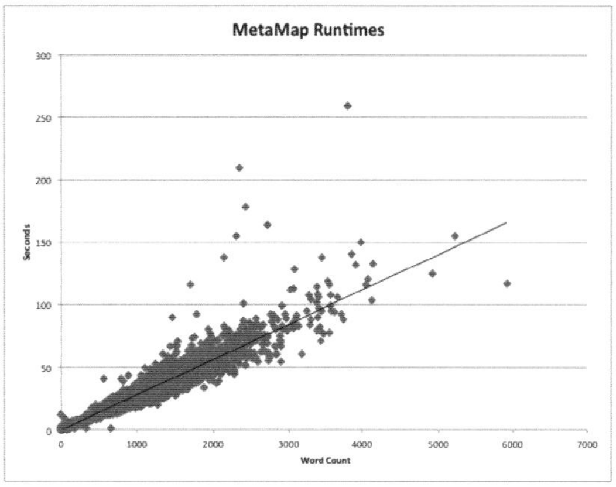

Figure 1: MetaMap Runtimes

1.3 Motivation for the resource: reproducibility

Any large-scale run of MetaMap over a huge document collection will have occasional failures, etc. The odds of any two runs having the same output are therefore slim. Moreover, there is potential variability in how documents are preprocessed for use with MetaMap. Using this repository of MetaMap outputs will ensure reproducibility of experiments and also preclude the necessity of performing the same preparatory work and MetaMap processing on the same data.

1.4 Motivation for the resource: semantic types

The creation of the MIMIC-II repository is an intermediate step in our research. We are extracting the semantic types found in each clinical note in an attempt to determine if there exists evidence of subdomains across the categories used by MIMIC-II to group the notes.

2 Materials and Methods

2.1 Materials

MetaMap is a program developed at the National Library of Medicine (NLM) that maps biomedical concepts to the UMLS Metathesaurus and reports on the corresponding semantic types.[1] The program is used extensively by researchers in the field of biomedical text mining. See Aronson, 2001; Aronson and Lang, 2010.

[1]Users of MetaMap must comply with the UMLS Metathesaurus license agreement (`https://uts.nlm.nih.gov/license.html`).

Although our focus is on the clinical notes contained in a single table, **noteevents**, MIMIC-II is both a relational database (PostgreSQL 9.1.9) containing 39 tables of clinical data and bedside monitor waveforms and the associated derived parameters and events stored in flat binary files (with ASCII header descriptors). For each Intensive Care Unit (ICU) patient, Saeed et al. (2011) collected a wide range of data including *inter alia* laboratory data, therapeutic intervention profiles, MD and nursing progress notes, discharge summaries, radiology reports, International Classification of Diseases, 9th Revision codes, and, for a subset of patients, high-resolution vital sign trends and waveforms. All data were scrubbed for personal information to ensure compliance with the Health Insurance Portability and Accountability Act (HIPAA). These data were then uploaded to a relational database thereby allowing for easy access to extensive information for each patient's stay in the ICU (Saeed et al.). A more detailed description of the use of the MIMIC-II database may be found in(Clifford et al. , 2012).

The abbreviated schema in Table 1 below shows that each ID uniquely identifies a note along with a SubjectID, a Category and Text.[2] We added the ID attribute to **noteevents** as a primary key because SubjectID, Category and Text are not keys. Thus, a particular patient might have many notes and many categories but each note is uniquely identified.

Attribute	Type	Cardinality	Sample Values
ID	integer	unique	1, 2, 3, 4...
SubjectID	integer	many to one ID	95, 100, 100, 99,
Category	character varying(26)	many to one ID	radiology
Text	text	many to one ID	interval placement of ICD

Table 1: Schema MIMIC-II noteevents table.

As mentioned above, the notes in the MIMIC-II database are categorized as MD reports, radiology reports, discharge summaries and nursing/other reports. The contents of these notes varied greatly. The MD and nursing notes tended to be short and unstructured with a number of abbreviations and misspellings, whereas the radiology reports were longer, more structured and showed fewer errors.

The MIMIC-II version used for this research is 2.6 (April 2011; 32,536 subjects).

The distribution of reports with summary statistics is below in Table 2.

[2]The full **noteevents** table has ten other attributes such as admission date, various timestamps and patient information but these were not relevant to our research.

	MD	Discharge	Radiology	Nursing	Totals
min words	0	0	0	0	0
max words	632	6684	2760	632	6684
median words	108	963.5	174	108	135
average words	131.6	1009.2	265.5	131.6	194.7
total notes	23,270	31,877	383,701	798,838	1,237,686

Table 2: MIMIC-II Clinical Note Summary Stats.

2.2 Methods

We used MetaMap to process the clinical notes in order to find semantic concepts, the latter of which are being used in our current research. For the work in this paper, we used MetaMap 2013 with the 2013AB database.

Before processing the notes with MetaMap, a number of preparatory steps were taken. As mentioned above, a primary key was added to the **noteevents** table to provide a unique id for each note. A Python script then queried the database extracting each note and storing it in a file named according to the following convention: uniqueID_subjectID_category.txt where uniqueID is the primary key value from the **noteevents** table, subjectID is the unique number assigned to each patient and category is one of the four categories mentioned above.

Each of the notes was then processed by a Bash shell script to remove blank lines and control characters. (This important step was added after a significant amount of processing had already taken place. If this is not done, a number of problems arise when running MetaMap). Finally, all files with 0 bytes were removed. These files were present because many tuples in the **noteevents** table contained clinical note entries with no data.

The number of options available when running MetaMap is considerable so we chose those that would provide a full and robust result set which would be useful to a wide range of researchers. In our first run, we limited the threshold for the Candidate Score to 1,000. However, for the repository, no threshold was set so that a full range of output is provided. [3]

The output is in XML in order to structure the data systematically and provide an easier and consistent way to parse the data. Although we chose XML initially, we intend to provide the same data in plain text and Prolog formats, again to provide utility to a broad range of researchers.

In order to process all files, a Bash shell script was created that called MetaMap on each note and created a corresponding XML file, named according to the same convention as that for notes but with the txt extension replaced by xml.

3 Results

3.1 The repository of MetaMap output

The repository for the MetaMap output contains an XML file for each note that originally contained text in the MIMIC-II database. Each XML file contains a wealth of information about each note and a discussion of this is beyond the scope of this paper (see `http://metamap.nlm.nih.gov/Docs/MM12_XML_Info.shtml`).

For our research, we are interested in the semantic types associated with phrases identified by MetaMap. Below is a section from output file 768591_19458_discharge.xml. This is a discharge summary for subject 19458 with a unique note id of 768591. The note contained the phrase "Admission Date" which MetaMap matched with a candidate score of 1000 and indicated that it is a temporal concept (tmco).

Ultimately, the MetaMap output files will be uploaded to the PhysioNet website and made available to the public.[4] The files will be organized in a fashion similar to the original data files on the site. Namely, data are grouped by subject ids and compressed in archives with approximately 1000 files each.

```
<Candidate>
<CandidateScore>-1000</CandidateScore>
<CandidateCUI>C1302393</CandidateCUI>
<CandidateMatched>Admission date
     </CandidateMatched>
<CandidatePreferred>Date of admission
     </CandidatePreferred>
<MatchedWords Count="2">
<MatchedWord>admission</MatchedWord>
<MatchedWord>date</MatchedWord>
</MatchedWords>
<SemTypes Count="1">
<SemType>tmco</SemType>
</SemTypes>
```

The original note contained 975 lines, whereas the MetaMap xml file contained 248,198. Thus it is obvious that there is a very large amount of MetaMap output that we don't consider but which may be of interest to other researchers.

[3]The exact MetaMap command we used was metamap13 –XMLf –silent –blanklines 3 filename.txt

[4]Subject again to the data usage agreement.

3.2 A Python module for manipulating MetaMap output

In order to make information in the XML files accessible to others, we developed a Python module (parseMM_xml.py) containing a number of methods or functions that allow one to parse the XML tree and extract relevant information.

Although we will add more functionality as needed and requested, at this point the following methods are implemented:

- parseXMLtree(filename) – parses the contents of filename and returns a node representing the top of the document tree.
- getXMLsummary(XMLtree) – summarizes the data contained in the parsed XML tree. The summary contains top-level elements and their corresponding text. The output is much like that contained in typical MetaMap text output.
- getCUIs(XMLtree) – returns the MetaMap CUIs found in the XML tree along with the matching concepts.
- getNegatedConcepts(XMLtree) – returns negated concepts and their corresponding CUIs.
- getSemanticTypes(XMLtree) – returns matched concepts, their CUIs, the candidate scores and the semantic types associated with the concept.
- findAttribute(attribute) – searches the document tree for an attribute of the user's choosing. Returns the attributes with their corresponding text values.

We chose Python to create our module because of its ease of use and its multi-platform capabilities. Once Python is installed and the parseMM_xml.py is placed in a directory along with the MetaMap xml file which is to be analyzed, retrieving relevant information is relatively straightforward. [5]

[5]Under most circumstances, Python is already installed on the Mac OS X and Linux operating systems.

A stylized version of our code is presented below.

```
# Parse XML tree and return semantic
   types.

import parseMM_xml
xml_tree = \
parseXMLtree("noteid_subid_category.xml")
semTypes = getSemanticTypes(xml_tree)

print(semTypes)
```

A truncated listing of the output:

CandidateCUI – C0011008
CandidateMatched – Date
1 – SemType – Temporal Concept
CandidateCUI – C2348077
CandidateMatched – Date
2 – SemType – Food

In order to fully test the robustness of our module, we will do further unit and regression testing, in addition to providing more exception handling. Ultimately, the code will be available on SourceForge, an Open Source web source code repository available at `www.sourceforge.net`.

Acknowledgments

We would like to thank George Moody of MIT for his help with questions concerning the MIMIC-II database.

References

Alan R. Aronson 2001. *Effective mapping of biomedical text to the UMLS Metathesaurus: the MetaMap program,* Proc AMIA Symp. 2001; 17:21.

Alan R. Aronson, Franois-Michel Lang 2010. *An overview of MetaMap: historical perspective and recent advances,* J Am Med Inform Assoc 2010; 17:3 229-236

Gari D. Clifford, Daniel J. Scott, Mauricio Villarroel 2012. *User Guide and Documentation for the MIMIC II Database,* `http://mimic.physionet.org/UserGuide/UserGuide.pdf`

Ary L. Goldberger; Luis A. N. Amaral; Leon Glass; Jeffrey M. Hausdorff; Plamen Ch. Ivanov; Roger G. Mark; Joseph E. Mietus; George B. Moody; Chung-Kang Peng; H. Eugene Stanley, 2000 *PhysioBank, PhysioToolkit, and PhysioNet: Components of a New Research Resource for Complex Physiologic Signals,* Circulation 101(23): e215-e220.

Mohammed Saeed, Mauricio Villarroel, Andrew T. Reisner, Gari Clifford, Li-Wei Lehman, George Moody, Thomas Heldt, Tin H. Kyaw, Benjamin Moody, Roger G. Mark. 2011 *The Multiparameter intelligent monitoring in intensive care II (MIMIC-II): A public-access ICU database,* Critical Care Medicine; 39(5):952-960

MetaMap Release Notes Website 2013. *MetaMap 2013 Release Notes* http://metamap.nlm.nih.gov/Docs/MM12_XML_Info.shtml

MetaMap 2012 Output Website 2014. *MetaMap 2012 XML Output Explained* http://metamap.nlm.nih.gov/Docs/MM12_XML_Info.shtml

PhysioNet Website 2014. *PhysioNet MIMIC-II Website* http://physionet.org/mimic2/

Extracting drug indications and adverse drug reactions from Spanish health social media

Isabel Segura-Bedmar, Santiago de la Peña, Paloma Martínez
Computer Science Department
Carlos III University of Madrid, Spain
{isegura|spena|pmf}@inf.uc3m.es

Abstract

In this paper, we present preliminary results obtained using a system based on co-occurrence of drug-effect pairs as a first step in the study of detecting adverse drug reactions and drug indications from social media texts. To the best of our knowledge, this is the first work that extracts this kind of relationships from user messages that were collected from an online Spanish health-forum. In addition, we also describe the automatic construction of the first Spanish database for drug indications and adverse drug reactions.

1 Introduction

The activity of Pharmacovigilance (science devoted to the detection and prevention of any possible drug-related problem, including adverse drug effects) has gained significant importance in the recent decades, due to the growing number of drug safety incidents (Bond and Raehl, 2006) as well as to their high associated costs (van Der Hooft et al., 2006).

Nowadays, the major medicine regulatory agencies such as the US Food and Drug Administration (FDA) or the European Medicines Agency (EMA) are working to create policies and practices to facilitate the reporting of adverse drug reactions (ADRs) by healthcare professionals and patients. However, several studies have shown that ADRs are under-estimated because many healthcare professionals do not have enough time to use the ADR reporting systems (Bates et al., 2003; van Der Hooft et al., 2006; McClellan, 2007) . In addition, healthcare professionals tend to report only those ADRs on which they have absolute certainty of their existence. Unlike reports from healthcare professionals, patient reports often provide more detailed and explicit information about ADRs (Herxheimer et al., 2010). Nevertheless, the rate of ADRs reported by patients is still very low probably because many patients are still unaware of the existence of ADR reporting systems. In addition, patients may feel embarrassed when describing their symptoms.

In this paper, we pose the hypothesis that health-related social media can be used as a complementary data source to the ADR reporting systems. In particular, health forums contain a large number of comments describing patient experiences that would be a fertile source of data to detect unknown ADRs.

Several systems have been developed for extracting ADRs from social media (Leaman et al., 2010; Nikfarjam and Gonzalez, 2011). However to the best of our knowledge, only one work in the literature has focused on the detection of ADRs from social media in Spanish (Segura-Bedmar et al., 2014). Indeed, it is only concerned with the detection of mentions of drugs and their effects, without dealing with the extraction of the relationships between them. In this paper, we extend this existing work in order to extract drug indications and adverse drug reactions from user comments in a Spanish health-forum.

The remaining of this paper is structured as follows: the next section surveys related work on ADR detection from social media. Section 3 describes the creation of a gold-standard corpus we used for our experiments. Sections 4 and 5 respectively describe the techniques employed and their results. Lastly, some conclusive remarks and future perspectives are given in Section 6.

2 Related Work

In recent years, the application of Natural Language Processing (NLP) techniques to mine drug indications and adverse drug reactions from texts has been explored with promising results, mainly in the context of drug labels (Gurulingappa et al.,

Proceedings of the 2014 Workshop on Biomedical Natural Language Processing (BioNLP 2014), pages 98–106,
Baltimore, Maryland USA, June 26-27 2014.

2013; Li et al., 2013; Kuhn et al., 2010; Fung et al., 2013), biomedical literature (Xu and Wang, 2013), medical case reports (Gurulingappa et al., 2012) and health records (Friedman, 2009; Sohn et al., 2011). However, as it will be described below, the extraction of these drug relationships from social media has received much less attention.

To date, most of research on drug name recognition concerns either biomedical literature (Segura-Bedmar et al., 2013; Krallinger et al., 2013) or clinical records (Uzuner et al., 2010), thus leaving unexplored this task in social media texts.

To our knowledge, there is no work in the literature that addresses the extraction of drug indications from social media texts. Regarding the detection of ADRs, Leaman et al., (2010) developed a system to automatically recognize adverse effects in user comments from the DailyStrength[1] health-related social network. A corpus of 3,600 comments was manually annotated with a total of 1,866 drug conditions, including beneficial effects, adverse effects, indications and others. This study focused only on a set of four drugs, and thereby, drug name recognition was not addressed. The system used a dictionary-based approach to identify adverse effects and a set of keywords in order to distinguish adverse effects from the other drug conditions. The dictionary consisted of 4,201 concepts, which were collected from several resources such as the COSTART vocabulary (FDA, 1970), the SIDER database (Kuhn et al., 2010), the MedEffect database [2] and a list of colloquial phrases manually collected from the comments. The system achieved a precision of 78.3% and a recall of 69.9% (an f-measure of 73.9%).

Later, Nikfarjam and Gonzalez (2011) applied association rule mining to extract frequent patterns describing opinions about drugs. The rules were generated using the Apriori tool, an implementation of the Apriori algorithm (Agrawal et al., 1994) for association rule mining. The main advantage of this approach over the dictionary based approach is that the system is able to detect terms not included in the dictionary. The results of this study were 70.01% precision and 66.32% recall, for an f-measure of 67.96%.

Benton et al.,(2011) collected a lexicon of lay medical terms from websites and databases about drugs and their adverse effects to identify drug effects. Then, the authors applied the Fishers exact test (Fisher, 1922) to find all the drug-effect pairs that co-occurred independently by chance in a corpus of user comments. To evaluate the system, the authors focused only on the four most commonly used drugs to treat breast cancer. Precision and recall were calculated by comparing the adverse effects from their drug labels and the adverse effects obtained by the system. The system obtained an average precision of 77% and an average recall of 35.1% for all four drugs.

To the best of our knowledge, the system described in (Segura-Bedmar et al., 2014) is the only one that has dealt with the detection of drugs and their effects from Spanish social media streams. The system used the Textalytics tool[3], which follows a dictionary-based approach to identify entities in texts. The dictionary was constructed based on the following resources: CIMA[4] and MedDRA[5]. CIMA is an online information center maintained by the Spanish Agency for Medicines and Health Products (AEMPS). CIMA provides information on all drugs authorized in Spain, though it does not include drugs approved only in Latin America. CIMA contains a total of 16,418 brand drugs and 2,228 generic drugs. Many brand drugs have very long names because they include additional information such as dosages, mode and route of administration, laboratory, among others (for example, *ESPIDIFEN 400 mg GRANULADO PARA SOLUCION ORAL SABOR ALBARICOQUE*). For this reason, brand drug names were simplified before being included in the dictionary. After removing the additional information, the resulting list of brand drug names consisted of 3,662 terms. Thus, the dictionary contained a total of 5,890 drugs. As regards to the effects, the authors decided to use MedDRA, a medical multilingual terminology dictionary about events associated with drugs. MedDRA is composed of a five levels hierarchy. A total of 72,072 terms from the most specific level, "Lowest Level Terms" (LLTs), were integrated into the dictionary. In addition, several gazetteers including drugs and effects were collected from websites such as Vademecum[6], a Spanish online website that provides information to patients on drugs and their side effects, and

[1]http://www.dailystrength.org/

[2]http://www.hc-sc.gc.ca/dhp-mps/medeff/index-eng.php

[3]https://textalytics.com/

[4]http://www.aemps.gob.es/cima/

[5]http://www.meddra.org/

[6]http://www.vademecum.es/

the ATC system[7], a classification system of drugs. Thus, the dictionary and the two gazetteers contained a total of 7,593 drugs and 74,865 effects. The system yielded a precision of 87% for drugs and 85% for effects, and a recall of 80% for drugs and 56% for effects.

3 The SpanishADR corpus

Segura-Bedmar et al., (2014) created the first Spanish corpus of user comments annotated with drugs and their effects. The corpus consists of 400 comments, which were gathered from ForumClinic[8], an interactive health social platform, where patients exchange information about their diseases and their treatments. The texts were manually annotated by two annotators with expertise in Pharmacovigilance. All the mentions of drugs and effects were annotated, even those containing spelling or grammatical errors (for example, *hemorrajia* (haemorrhage)). An assessment of the inter-annotator agreement (IAA) was based on the F-measure metric, which approximates the kappa coefficient (Cohen, 1960) when the number of true negatives (TN) is very large (Hripcsak and Rothschild, 2005). This assessment revealed that while drugs showed a high IAA (0.89), their effects point to moderate agreement (0.59). This may be due to drugs have specific names and there are a limited number of them, however their effects are expressed by patients in many different ways due to the variability and richness of natural language. The corpus is available for academic purposes[9].

In this paper, we extend the Spanish corpus to incorporate the annotation of the relationships between drugs and their effects. In particular, we annotated drug indications and adverse drug reactions. These relationships were annotated at comment level rather than sentence level, because determining sentence boundaries in this kind of texts can be problematic since many users often write ungrammatical sentences. Guidelines were created by two annotators (A1, A2) and a third annotator (A3) was trained on the annotation guidelines. Then, we split the corpus in three subsets, and each subset was annotated by one annotator. Finally, IAA was measured using kappa-statistic on a sample of 97 documents randomly selected. These documents were annotated by the three annotators and annotation differences were analysed.

As Table 1 shows, the resulting corpus has 61 drug indications and 103 adverse drug reactions. The average size of a comment is 72 tokens. The average size of a text fragment describing a drug indication is 34.7 tokens and 28.2 tokens for adverse drug reactions.

Annotation	**Size**
drugs	188
effect	545
drug indication	61
adverse drug reaction	103

Table 1: Size of the extended SpanishADR corpus.

As it is shown in Table 2, the IAA figures clearly suggest that the annotators have high agreement among them. We think that the IAA figures were lower with the third annotator because he did not participate in the guidelines development process, and maybe, he was not trained well enough to perform the task. The main source of disagreement among the annotators could arise from considering whether a term refers to a drug effect or not. This is due to some terms are too general (such as *trastorno* (upset), *enfermedad* (disease), *molestia* (ache)). The annotators A1 and A2, in general, ruled out all the relation instances where these general terms occur, however they were considered and annotated by the third annotator.

	A2	**A3**
A1	0.8	0.69
A2	-	0.68

Table 2: Pairwise IAA for each combination of two annotators. IAA was measured using Cohens' kappa statistic

4 Methods

In this contribution, some refinements to the system (Segura-Bedmar et al., 2014) are proposed. The error analysis performed in (Segura-Bedmar et al., 2014) showed that most of false positives for drug effects were mainly due to the inclusion of MedDRA terms referring to procedures and tests in the dictionary. MedDRA includes terms for diseases, signs, abnormalities, procedures and tests. Therefore, we decided not to include terms corresponding to the "Procedimientos

[7]http://www.whocc.no/atc_ddd_index/
[8]http://www.forumclinic.org/
[9]http://labda.inf.uc3m.es/SpanishADRCorpus

médicos y quirúrgicos" and "Exploraciones complementarias" categories since they do not represent drug effects. Thus, we created a new dictionary that only includes those terms from MedDRA that actually refer to drug effects. As in the system (Segura-Bedmar et al., 2014), we applied the Textalytics tool, which follows a dictionary-based approach, to identify drugs and their effects occurring in the messages. We created a GATE[10] pipeline application integrating the Textalytic module and the gazetteers collected from the Vademecum website and the ATC system proposed in (Segura-Bedmar et al., 2014).

In addition, we created an additional gazetteer in order to increase the coverage. We developed a web crawler to browse and download pages related to drugs from the MedLinePlus website[11]. Unlike Vademecum, which only contains information for drugs approved in Spain, MedLinePlus also includes information about drugs only approved in Latin America. Terms describing drug effects were extracted by regular expressions from these pages and then were incorporated into a gazetteer. Then, the new gazetteer was also integrated into the GATE pipeline application to identify drugs and effects. Several experiments with different settings of this pipeline are described in the following section.

The main contribution of this paper is to propose an approach for detecting relationships between drugs and their effects from user comments in Spanish. The main difficulty in this task is that although there are several English databases such as SIDER or MedEffect with information about drugs and their side effects, none of them are available for Spanish. Moreover, these resources do not include drug indications. Thus, we have automatically built the first database, *SpanishDrugEffectBD*, with information about drugs, their drug indications as well as their adverse drug reactions in Spanish. Our first step was to populate the database with all drugs and effects from our dictionary. Figure 1 shows the database schema. Active ingredients are saved into the *Drug* table, and their synonyms and brand names into the *DrugSynset* table. Likewise, concepts from MedDRA are saved into the *Effect* table and their synonyms are saved into the *EffectSynset* table. As it is shown in Figure 1, the database is also designed to store external ids from other databases. Thus, drugs and effects can be linked to external databases by the tables *has_externalIDDrug* and *has_externalIDDrug*, respectively.

To obtain the relationships between drugs and their effects, we developed several web crawlers in order to gather sections describing drug indications and adverse drug reactions from drug package leaflets contained in the following websites: MedLinePlus, Prospectos.Net[12] and Prospectos.org[13]. Once these sections were downloaded, their texts were processed using the TextAlyticis tool to recognize drugs and their effects. As each section (describing drug indications or adverse drug effects) is linked to one drug, we decided to consider the effects contained in the section as possible relationships with this drug. The type of relationship depends on the type of section: drug indication or adverse drug reaction. Thus for example, a pair (drug, effect) from a section describing drug indications is saved into the *DrugEffect* table as a drug indication relationship, while if the pair is obtained from a section describing adverse drug reactions, then it is saved as an adverse drug reaction. This database can be used to automatically identify drug indications and adverse drug reactions from texts. Table 3 shows the number of drugs, effects and their relationships stored into the database.

	Concepts	Synonyms
drugs	3,244	7,378
effects	16,940	52,199
drug indications	4,877	
adverse drug reactions	58,633	

Table 3: Number of drugs, effects, drug indications and adverse drug effects in the SpanishDrugEffectBD database.

As regards to the extraction of the relationships between drugs and their effects occurring in the corpus, first of all, texts were automatically annotated with drugs and effects using the GATE pipeline application. Then, in order to generate all possible relation instances between drugs and their effects, we considered several sizes of window: 10, 20, 30, 40 and 50. Given a size n, any pair (drug, effect) co-occurring within a window of n-tokens are treated as a relation instance. Af-

[10]http://gate.ac.uk/

[11]http://www.nlm.nih.gov/medlineplus/spanish/

[12]http://www.prospectos.net/

[13]http://prospectos.org/

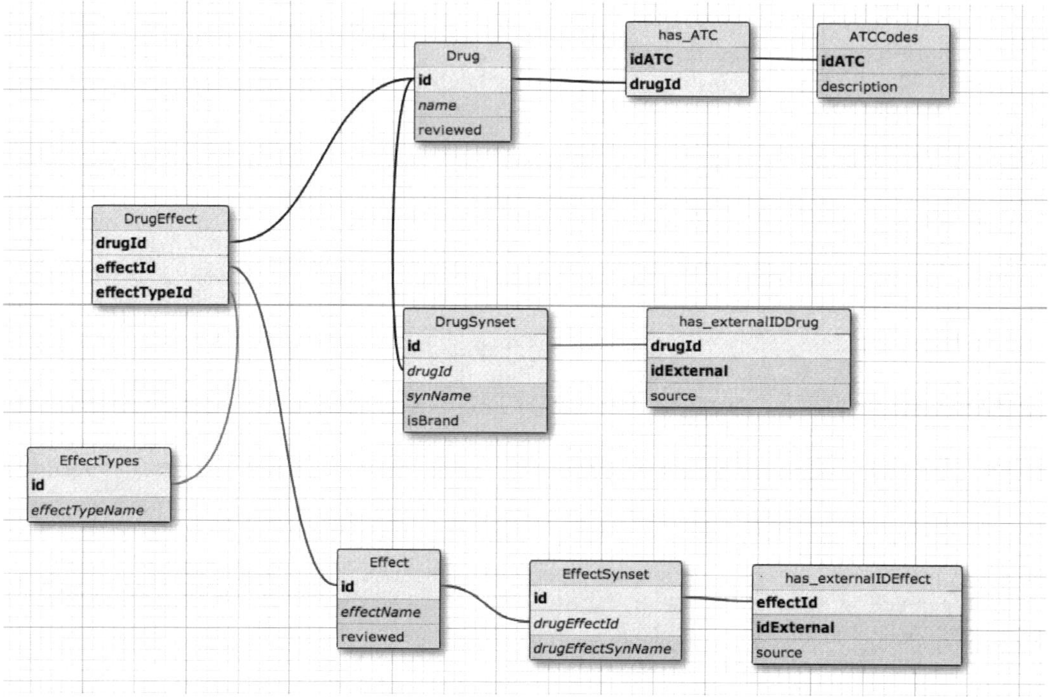

Figure 1: The SpanishDrugEffectBD database schema

terwards, each relation instance is looked up in the *DrugEffect* table in order to determine if it is a positive instance and if this is the case, its type: drug indication or adverse drug reaction.

5 Experiments

Several experiments have been performed in order to evaluate the contribution of the proposed methods and resources. Table 4 shows the results for the named entity recognition task of drugs and effects using the dictionary integrated into the TextAlytic tool. The first row shows the results with the dictionary built from the CIMA and MedDRA resources, while the second one shows the results obtained using the new dictionary in which those MedDRA terms corresponding to "Procedimientos médicos y quirúrgicos" and "Exploraciones complementarias" categories were ruled out. As it can be seen in this table, the new dictionary permits to obtain a significant improvement with respect to the original dictionary. For effect type, precision was increased almost a 40% and recall a 7%. As regards to the contribution of the gazetteers, the coverage for effects improves almost a 6% but with significant decrease in precision of almost 21%. Regarding to the detection of drugs, the use of gazetteers improves slightly the precision and achieves a significant improvement in the recall of almost 35%.

The major cause of false negatives for drug effects was the use of colloquial expressions (such as *'me deja ko'* (it makes me ko)) to describe an adverse effect. These phrases are not included in our dictionary. Another important cause was the dictionary and gazetteers do not cover all the lexical variations of a same effect (for example *depresión* (depression), *depresivo* (depress), *me deprimo* (I get depressed)). In addition, many false negatives were due to spelling mistakes (for example *hemorrajia* instead of *hemorragia* (haemorrhage)) and abbreviations (*depre* is an abbreviation for *depresión* (depression)).

Regarding to the results for the relation extraction task, Table 5 shows the overall results obtained using a baseline system, which considers all pairs (drug, effect) occurring in messages as positive relation instances, and a second approach using the SpanishDrugEffectBD database (a relation instance is positive only if it is found into the database). In both experiments, a window size of 250 tokens was used. The database provides a high precision but with a very low recall of only 15%.

Approach	Entity	P	R	F1
Dictionary	drugs	0.84	0.46	0.60
	effect	0.45	0.38	0.41
New dictionary	drugs	0.84	0.46	0.60
	effect	0.84	0.45	0.59
New dictionary plus gazetteers	drugs	0.86	0.81	0.84
	effect	0.63	0.51	0.57

Table 4: Precision, Recall and F-measure for named entity recognition task.

As it can be seen in Table 6, when the type of the relationship is considered, the performance is even lower.

Approach	P	R	F1
Baseline	0.31	1.00	0.47
SpanishDrugEffectBD	0.83	0.15	0.25

Table 5: Overall results for relation extraction task (window size of 250 tokens).

Relation	P	R	F1
Drug indication	0.50	0.02	0.03
Adverse drug reaction	0.65	0.11	0.18

Table 6: Results for drug indications and adverse drug reactions using only the database (window size of 50 tokens).

Figure 2 shows an example of the output of our system using the database. The system is able to detect the relationship of indication between *alprazolman* and *ansiedad* (anxiety), but fails in detecting the adverse drug reaction between *alprazolman* and *dependencia* (dependency). The adverse drug reaction between *lamotrigina* and *vertigo* is detected.

The co-occurrence approach provides better results than the use of the database. Table 7 shows the results for different size of windows. As it was expected, small sizes provide better precision but lower recall.

Size of window	P	R	F1
10	0.71	0.24	0.36
20	0.59	0.53	0.56
30	0.52	0.69	0.59
40	0.47	0.77	0.58
50	0.44	0.84	0.58

Table 7: Overall results for relation extraction task using the co-occurrence approach considering different window sizes.

6 Conclusion

In this paper we present the first corpus where 400 user messages from a Spanish health social network have been annotated with drug indications and adverse drug reactions. In addition, we present preliminary results obtained using a very simple system based on co-occurrence of drug-effect pairs as a first step in the study of detecting adverse drug reactions and drug indications from social media streams. Results show that there is still much room for improvement in the identification of drugs and effects, as well as in the extraction of drug indications and adverse drug rections.

As it was already mentioned in Section 2, the recognition of drugs in social media texts has hardly been tackled since most systems were focused on a given and fixed set of drugs. Moreover, little research has been conducted to extract relationships between drugs and their effects from social media. Most systems for extracting ADRs follow a dictionary-based approach. The main drawback of these systems is that they fail to recognize terms which are not included in the dictionary. In addition, the dictionary-based approach is not able to handle the large number of spelling and grammar errors in social media texts. Moreover, the detection of ADRs and drug indications has not been attempted for languages other than English. Indeed, automatic information extraction from Spanish-language social media in the field of health remains largely unexplored.

Social media texts pose additional challenges to those associated with the processing of clinical records and medical literature. These new challenges include the management of meta-information included in the text (for example as tags in tweets)(Bouillot et al., 2013), the detection of typos and unconventional spelling, word short-

Results - Drugs: Indications and Adverse Effects

Text Annotated

DEPR357 segun muchos prospectos no se deben pasar 21 dias (3 semanas). . el alprazolam es precisamente el ansiolitico que mas dependencia genera por su alta potencia y vida media corta. . pero si tuviera que darte mi opinion, te diria que depende de la "adiccion a la ansiedad" que tengas. si andas medio tranquilo y bien, la dependencia no es tanta y los puedes soltar con facilidad; si tu ansiedad es altisima no los vas a poder dejar. . un saludo vecino

TRBI11122 hola anais. . finalmente consegui hablar con mi psiquiatra habitual y me dijo que el vertigo era un efecto secundario de la lamotrigina. le comente que lo habia suspendido. asi esta informado de mi situacion

Drug Effect Indication AdverseEffect

Figure 2: An example of the output of the system using the database.

enings (Neunerdt et al., 2013; Moreira et al., 2013) and slang and emoticons (Balahur, 2013), among others. Another challenge that should be taken into account is that while clinical records and medical literature can be mapped to terminological resources or biomedical ontologies, lay terminology used by patients to describe their treatments and their effects, in general, is not collected in any terminological resource, which would facilitate the automatic processing of this kind of texts.

In this paper, we also describe the automatic creation of a database for drug indications and adverse drug reactions from drug package leaflets. To the best of our knowledge, this is the first database available for Spanish. Although the use of this database did not improve the results due to its limited coverage, we think that the database could be a valuable resource for future efforts. Thus, we plan to translate the database into an ontology and to populate it with more entities and relationships. As future work, we plan the following tasks:

- To create a lexicon containing idiomatic expressions used by patients to express drug effects.
- To use techniques such as lemmatization and stemming to cope with the problem of lexical variability and to resolve abbreviations.
- To integrate advanced matching methods capable of dealing with the spelling error problem.
- To increase the size of the corpus.
- To apply a SVM classification approach to extract relationships between drugs and their effects.

We hope our research will be beneficial to AEMPS as well as to the pharmaceutical industry in the improvement of their pharmacovigilance systems. Both the corpus and the database are freely available online[14] for research purposes.

Acknowledgments

This work was supported by the EU project TrendMiner [FP7-ICT287863], by the project MULTIMEDICA [TIN2010-20644-C03-01], and by the Research Network MA2VICMR [S2009/TIC-1542].

References

Rakesh Agrawal, Ramakrishnan Srikant, et al. 1994. Fast algorithms for mining association rules. In *Proc. 20th int. conf. very large data bases, VLDB*, volume 1215, pages 487–499.

Alexandra Balahur. 2013. Sentiment analysis in social media texts. *WASSA 2013*, page 120.

David W Bates, R Scott Evans, Harvey Murff, Peter D Stetson, Lisa Pizziferri, and George Hripcsak. 2003. Detecting adverse events using information technology. *Journal of the American Medical Informatics Association*, 10(2):115–128.

Adrian Benton, Lyle Ungar, Shawndra Hill, Sean Hennessy, Jun Mao, Annie Chung, Charles E Leonard, and John H Holmes. 2011. Identifying potential adverse effects using the web: A new approach to

[14]http://labda.inf.uc3m.es/SpanishADRCorpus

medical hypothesis generation. *Journal of biomedical informatics*, 44(6):989–996.

CA Bond and Cynthia L Raehl. 2006. Adverse drug reactions in united states hospitals. *Pharmacotherapy: The Journal of Human Pharmacology and Drug Therapy*, 26(5):601–608.

Flavien Bouillot, Phan Nhat Hai, Nicolas Béchet, Sandra Bringay, Dino Ienco, Stan Matwin, Pascal Poncelet, Mathieu Roche, and Maguelonne Teisseire. 2013. How to extract relevant knowledge from tweets? In *Information Search, Integration and Personalization*, pages 111–120. Springer.

Jacob Cohen. 1960. A coefficient of agreement for nominal scales. *Educational and Psychological Measurement*, 20(1):37–46.

FDA. 1970. National adverse drug reaction directory: Costart (coding symbols for thesaurus of adverse reaction terms). *Rock-Irvine, Charles F, Sharp,) r, MD, Huntington Memorial Hospital, Stuart l, Silverman, MD, University of California, los Angeles, West los Angeles-Veterans Affairs Medical Center, Osteoporosis Medical Center.*

Ronald A Fisher. 1922. On the interpretation of chi-squared from contingency tables, and the calculation of p. *Journal of the Royal Statistical Society*, 85(1):87–94.

Carol Friedman. 2009. Discovering novel adverse drug events using natural language processing and mining of the electronic health record. In *Artificial Intelligence in Medicine*, pages 1–5. Springer.

Kin Wah Fung, Chiang S Jao, and Dina Demner-Fushman. 2013. Extracting drug indication information from structured product labels using natural language processing. *Journal of the American Medical Informatics Association*, 20(3):482–488.

Harsha Gurulingappa, Abdul Mateen-Rajput, Luca Toldo, et al. 2012. Extraction of potential adverse drug events from medical case reports. *J Biomed Semantics*, 3(1):15.

Harsha Gurulingappa, Luca Toldo, Abdul Mateen Rajput, Jan A Kors, Adel Taweel, and Yorki Tayrouz. 2013. Automatic detection of adverse events to predict drug label changes using text and data mining techniques. *Pharmacoepidemiology and drug safety*, 22(11):1189–1194.

A Herxheimer, MR Crombag, and TL Alves. 2010. Direct patient reporting of adverse drug reactions. a twelve-country survey & literature review. *Health Action International (HAI)(Europe). Amsterdam.*

George Hripcsak and Adam S Rothschild. 2005. Agreement, the f-measure, and reliability in information retrieval. *Journal of the American Medical Informatics Association*, 12(3):296–298.

Martin Krallinger, Florian Leitner, Obdulia Rabal, Miguel Vazquez, Julen Oyarzabal, and Alfonso Valencia. 2013. Overview of the chemical compound and drug name recognition (chemdner) task. In *BioCreative Challenge Evaluation Workshop vol. 2*, page 2.

Michael Kuhn, Monica Campillos, Ivica Letunic, Lars Juhl Jensen, and Peer Bork. 2010. A side effect resource to capture phenotypic effects of drugs. *Molecular systems biology*, 6(1).

Robert Leaman, Laura Wojtulewicz, Ryan Sullivan, Annie Skariah, Jian Yang, and Graciela Gonzalez. 2010. Towards internet-age pharmacovigilance: extracting adverse drug reactions from user posts to health-related social networks. In *Proceedings of the 2010 workshop on biomedical natural language processing*, pages 117–125. Association for Computational Linguistics.

Qi Li, Louise Deleger, Todd Lingren, Haijun Zhai, Megan Kaiser, Laura Stoutenborough, Anil G Jegga, Kevin Bretonnel Cohen, and Imre Solti. 2013. Mining fda drug labels for medical conditions. *BMC medical informatics and decision making*, 13(1):53.

Mark McClellan. 2007. Drug safety reform at the fdapendulum swing or systematic improvement? *New England Journal of Medicine*, 356(17):1700–1702.

Silvio Moreira, Joao Filgueiras, and Bruno Martins. 2013. Reaction: A naive machine learning approach for sentiment classification. In *Proceedings of the 7th InternationalWorkshop on Semantic Evaluation (SemEval 2013)*, page 490.

Melanie Neunerdt, Michael Reyer, and Rudolf Mathar. 2013. A pos tagger for social media texts trained on web comments. *Polibits*, 48:59–66.

Azadeh Nikfarjam and Graciela H Gonzalez. 2011. Pattern mining for extraction of mentions of adverse drug reactions from user comments. In *AMIA Annual Symposium Proceedings*, volume 2011, page 1019. American Medical Informatics Association.

Isabel Segura-Bedmar, Paloma Martínez, and Marıa Herrero-Zazo. 2013. Semeval-2013 task 9: Extraction of drug-drug interactions from biomedical texts (ddiextraction 2013). *Proceedings of Semeval*, pages 341–350.

Isabel Segura-Bedmar, Ricardo Revert, and Paloma Martnez. 2014. Detecting drugs and adverse events from spanish social media streams. In *Proceedings of the 5th International Louhi Workshop on Health Document Text Mining and Information Analysis (Louhi 2014).*

Sunghwan Sohn, Jean-Pierre A Kocher, Christopher G Chute, and Guergana K Savova. 2011. Drug side effect extraction from clinical narratives of psychiatry and psychology patients. *Journal of the American Medical Informatics Association*, 18(Suppl 1):i144–i149.

Özlem Uzuner, Imre Solti, and Eithon Cadag. 2010. Extracting medication information from clinical text. *Journal of the American Medical Informatics Association*, 17(5):514–518.

Cornelis S van Der Hooft, Miriam CJM Sturkenboom, Kees van Grootheest, Herre J Kingma, and Bruno H Ch Stricker. 2006. Adverse drug reaction-related hospitalisations. *Drug Safety*, 29(2):161–168.

Rong Xu and QuanQiu Wang. 2013. Large-scale extraction of accurate drug-disease treatment pairs from biomedical literature for drug repurposing. *BMC bioinformatics*, 14(1):181.

Symptom recognition issue

Laure Martin
MoDyCo
Paris Ouest University
laure.martin.1988
@gmail.com

Delphine Battistelli
MoDyCo
Paris Ouest University
del.battistelli
@gmail.com

Thierry Charnois
LIPN
Paris 13 University
thierry.charnois
@lipn.univ-paris13.fr

Abstract

This work focuses on signs and symptoms recognition in biomedical texts abstracts. First, this specific task is described from a linguistic point of view. Then a methodology combining pattern mining and language processing is proposed. In the absence of an authoritative annotated corpus, our approach has the advantage of being weakly-supervised. Preliminary experimental results are discussed and reveal promising avenues.

1 Introduction

Our work is part of the Hybride[1] Project, which aims to expand the Orphanet encyclopedia. Orphanet is the reference portal for information on rare diseases (RD) and orphan drugs, for all audiences. A disease is considered rare if it affects less than 1 person in 2,000. There are between 6,000 and 8,000 RD. 30 million people are concerned in Europe. Among its activities, Orphanet maintains an RD encyclopedia by manually monitoring scientific publications. Hybride Project attempts to automatically acquire new RD-related knowledge from large amounts of scientific publications. The elements of knowledge about a disease are varied: onset, prevalence, signs and symptoms, transmission mode, disease causes (etiology).

In this article, we investigate the automatic recognition of signs and symptoms in abstracts from scientific articles. Although named entity recognition in the biomedical domain has been extensively studied, signs and symptoms seem to have been left aside, for there is very little work on the subject. First, the linguistic issue of our study is presented in section 2, then the state of the art and the description of our lexical resources in section 3. Then our corpus and general method are presented in section 4. First experiments are introduced in section 5. Finally, the work to come is presented in section 6.

[1]http://hybride.loria.fr/

2 Signs and symptoms

Signs and symptoms both refer to the features of a disease, except that a symptom (or functional sign) is noticed and described by a patient, whilst a clinical sign is observed by a healthcare professional. In thesauri and medical ontologies, these two notions are generally put together in the same category. Moreover, in texts –particularly in our corpus of abstracts from scientific articles– there is no morphological or syntactic difference between sign and symptom. The difference is only semantic, so it is impossible for non-specialists in the medical field to tell the difference from the linguistic context alone. In example (1), clinical signs are in bold and symptoms are italicized.

> (1) Cluster headache (CH) is a primary *headache* disease characterized by recurrent short-lasting attacks of excruciating unilateral periorbital *pain* accompanied by **ipsilateral autonomic signs** (*lacrimation*, **nasal congestion**, **ptosis**, **miosis**, lid **edema**, and eye **redness**).

Furthermore, the diagnosis is established by the symptoms and the clinical signs together. We did not, therefore, try to distinguish them.

Signs and symptoms take on the most varied linguistic forms, as is noticeable in the corpus (which will be described in more detail below). In its simplest form, a sign or symptom is a noun, which may be extended by complements, such as adjectives or other nouns (example 2). They also appear in other, more complex, forms, ranging from a single phrase to a whole sentence (example 3).

> (2) With disease progression patients additionally develop **weakness** and

Proceedings of the 2014 Workshop on Biomedical Natural Language Processing (BioNLP 2014), pages 107–111,
Baltimore, Maryland USA, June 26-27 2014.

wasting of the limb and bulbar muscles.

(3) Diagnosis is based on clinical presentation, and **glycemia and lactacidemia levels, after a meal (hyperglycemia and hypolactacidemia), and after three to four hour fasting (hypoglycemia and hyperlactacidemia)**.

In addition to their variety, the linguistic units representing signs and symptoms present some syntactic ambiguities, particularly ambiguities concerning prepositional attachment and coordination scope. In example (2), the first occurrence of "and" is ambiguous, for we don't know if "weakness" and "wasting" should be grouped together as a single manifestation of the disease, or if "weakness" on the one hand and "wasting of the limbs and bulbar muscles" on the other hand are two separate entities, as annotated here.

In addition to these syntactic ambiguities, two annotation difficulties also arise. The first one consists in correctly delimiting the linguistic units of the signs and symptoms (example 4a). We agreed with experts in the field that, generally, pieces of information such as adjectives of intensity or anatomical localizations were not part of the units; nevertheless, this information is interesting in that it provides the linguistic context for the signs and symptoms. The second difficulty concerns elliptical constructions: where two signs can be distinguished, only one can be annotated because the two nouns have an adjective in common (example 4b).

(4) In the severe forms, **paralysis** (4a) concerns the neck, shoulder, and proximal muscles, followed by involvement of the muscles of the distal upper extremities, the diaphragm and respiratory muscles, which may result in **respiratory compromise or arrest** (4b).

Eventually, the last difficulty that was met during the corpus observation is the semantic ambiguity existing between sign or symptom and disease denominations. A disease can be the clinical sign of another disease. A clinical sign may be included in a disease name or conversely. In example (5), the clinical sign is in bold and the name of the disease is underlined.

(5) The adult form results in progressive <u>limb-girdle **myopathy**</u> beginning with the lower limbs, and affects the respiratory system.

3 State of the art

Signs and symptoms have seldom been studied for themselves in the field of biomedical information extraction. They are often included in more general categories such as "clinical concepts" (Wagholikar et al., 2013), "medical problems" (Uzuner et al., 2011) or "phenotypic information" (South et al., 2009). Moreover, most of the studies are based on clinical reports or narrative corpora –the Mayo Clinic corpus (Savova et al., 2010) or the 2010i2b2/VA Challenge corpus (Uzuner et al., 2011)–, except for the Swedish MEDLEX Corpus (Kokkinakis, 2006), which comprises teaching material, guidelines, official documents, scientific articles from medical journals, etc. Our work aims at scientific monitoring and is therefore based on a corpus of abstracts from scientific articles.

Most of the information extraction systems developed in the works previously cited use lexical resources, such as the Unified Medical Language System (UMLS) or Medical Subject Headings (MeSH) thesauri for the named entity extraction task. The UMLS comprises over 160 controlled vocabularies such as MeSH, which is a generic medical thesaurus containing over 25,000 descriptors. However, as Albright et al. (2013) pointed out, UMLS was not originally designed for annotation, so some of the semantic types overlap. They add that "the sheer size of the UMLS schema increases the complexity of the annotation task and slows annotation, while only a small proportion of the annotation types present are used." That is why they decided to work with UMLS semantic groups instead of types, except for signs and symptoms –originally a semantic type in the Disorders semantic group–, that they used independently.

In a genetic disease context, a sign or symptom may be phenotype-related. A phenotype is all the observable characteristics of a person, such as their morphology, biochemical or physiological properties. It results from the interactions between a genotype (expression of an organism's genes) and its environment. As many rare diseases are genetic, many signs and symptoms may be found in lists of phenotype anomalies. For that reason, we chose to use the Human Phenotype Ontology – HPO (Khler et al., 2014) as a lexical resource. To our knowledge, HPO has not yet been used in any study on signs and symptoms extraction. Nevertheless, it should be recalled that phenotype anomalies are not always clinical signs, and signs or symptoms are not all phenotype-related. Even so, we decided to use HPO as a lexical resource because it lists 10,088 terms describing human phenotype anomalies and can be easily collected.

Just a very few studies take advantage of considering the linguistic contexts of sign and symptom entities. Kokkinakis (2006), after a first annotation step of his corpus with MeSH, states that 75% of the signs and symptoms co-occur with up to five other signs and symptoms in a sentence. This allowed him to develop new annotation rules. We can also mention the MedLEE system (Friedman, 1997), which provides, for each concept, its type (e.g. "problem"), value (e.g. "pain") and modifiers such as the degree (e.g. "severe") or the body location (e.g. "chest").

As far as we are concerned, our approach is based on the combination of NLP and pattern mining techniques. We will see that the linguistic contexts mentioned above are part of the patterns automatically discovered with our text mining tool.

4 Corpus and general method

As mentioned above, HPO was selected as the lexical resource for this project. With the list of phenotype anomalies as queries, we compiled a corpus of 306,606 abstracts from the MEDLINE database with the PubMed search engine. These abstracts are from articles published within the last 365 days. They consist of an ID, a title and a paragraph. Then, we applied HPO and kept only the sentences containing a unit annotated as a sign or symptom. As already pointed out, signs and symptoms are not all phenotype-related, so our pre-annotation is incomplete. Nonetheless, this first annotation is quick and cheap, and it initiates the process.

Figure 1 illustrates the successive steps in the approach. In step 1, HPO (f) is used to annotate a first corpus (a) by a single projection of HPO terms onto the texts. This annotated corpus provides a first learning corpus (b) to discover patterns (c) by a text mining method (step 2; this method is detailed below). These patterns are then validated by an expert (step 3), as linguistic patterns (d). Step

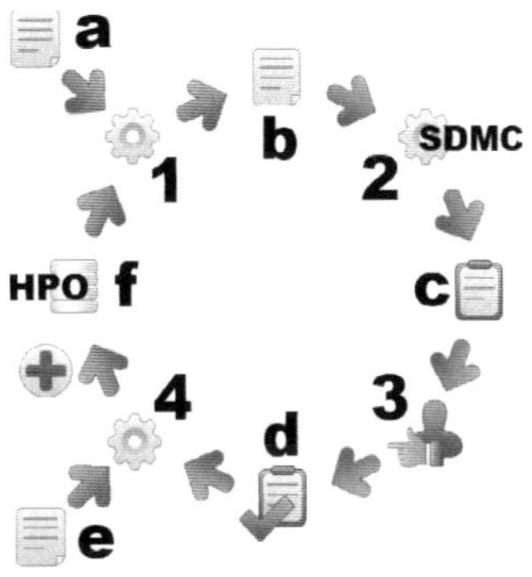

Figure 1: Iterative process of our sign and symptom extraction method

4 consists in using these patterns to annotate new corpora (e) and extract new terms (here with the semantic type of sign or symptom), which will be added to the resources (f). The process is finally repeated (back to step 1, with enriched lexical resources). This incremental process has the advantage of being weakly-supervised and non-dependent on the corpus type.

Sequential pattern mining was first introduced by Agrawal et al. (1995) in the data mining field. It was adapted to information extraction in texts by Béchet et al. (2012). It is a matter of locating, in a set of sequences, sequences of items having a frequency above a given threshold (called "support"). Pattern mining is done in an ordered sequence of items base, where each sequence corresponds to a text unit (the sentence here). An item represents a word in this sequence, generally the inflected form or the lemma or even the part of speech if the aim is to identify generic patterns. A number of parameters can be adapted along with the application.

Contrary to classical Machine Learning approaches which produce numerical models that are unintelligible for humans, data mining allows the discovery of symbolic patterns which can be interpreted by an expert. In the absence of authoritative annotated corpora for the recognition of signs and symptoms, manual validation of the patterns step is necessary, and often a large number of patterns still remains. To overcome this difficulty, Béchet et al. (2012) suggested adding constraints in order to reduce the results. In continuation of this work, we make use of the sequential patterns extraction tool SDMC[2], which makes it possible to

[2] https://sdmc.greyc.fr/

apply various constraints and condensed representations extraction (patterns without redundancy).

We adapted pattern mining to our field of application. Thus we first propose to use TreeTagger (Schmidt, 1994) as a pretreatment, in order to mark up different types of item (inflected form, lemma, part of speech). To narrow down the number of patterns returned by the tool, we introduce several constraints specific to our application: linguistic *membership* constraints (for example, we can choose to return only patterns containing at least one sign or symptom name), or the "gap" constraint (Dong and Pei, 2007), corresponding to possible gaps between items in the pattern. Thus a gap of maximal value n means that at most n items (words) are between each item of the pattern in the corresponding sequences (sentences).

5 First experiment

Annotating the first MEDLINE corpus of Abstracts with HPO provided us with a corpus of 10,000 annotated sentences. The 13,477 annotated units were replaced by a keyword –SYMPTOM– in order to facilitate the discovery of patterns. Then we used SDMC to mine the corpus for maximal patterns, with a minimal support of 10, a length between 3 and 50 words and a gap constraint of $g(0,0)$, i.e. the words are consecutive (no gap allowed). We were mining for lemma sequences only.

Results produced 988 patterns, among which 326 contained the keyword symptom. Based on these patterns, several remarks can already be made:

- Several annotated signs or symptoms are regularly associated with a third term, which can be another sign or symptom: {symptom}{symptom}{and}{stress};
- HPO annotation limitations (see section 3) are made visible by some contexts: {disease}{such}{as}{symptom};
- Some contexts are particularly recurrent, such as {be}{associate}{with}{symptom} or {characterize}{by}{symptom};
- Some temporal and chronological ordering contexts are present: {@card@}{%}{follow}{by}{symptom};
- The term "patient" is quite regular ({patient}{have}{severe}{symptom}), but after the evaluation, these occurrences turned out to be disease-related more than sign or symptom-related;
- The body location proved to be another regular context: {frontotemporal}{symptom}{ftd}.

Firstly, a linguistics expert selected the patterns that he considered the most relevant. These patterns were then classified in three categories: strong if they seem to strongly imply the presence of signs and symptoms (43 patterns), moderate (309 patterns) and weak (45 patterns). Secondly, these patterns were applied on a new corpus of MEDLINE abstracts in order to annotate the sign and symptom contexts. For the moment, only strong patterns have been applied.

25 abstracts were randomly selected among all the scientific articles published within the last month and dealing with Pompe disease. These 25 articles were manually annotated for signs and symptoms by an expert and thus constituted a gold standard. Then, we compared the manual annotation to our automatically annotated contexts. If the annotated sentence includes signs or symptoms, we consider that the annotation is relevant. Among the 25 abstracts (225 sentences), 27 contexts were extracted with our method. 23 were correct, 4 were irrelevant; 70 sentences were not annotated by the system. Thus the results were 23.7 in recall, reaching 82.2 in precision (36.8 in F-score).

6 Conclusions

Sign/disease ambiguity is the cause of 3 of the 4 irrelevant annotations, i.e. diseases were in the same linguistic context than signs. Thus the sentences were annotated but they contained diseases, not signs. The forth irrelevant annotation indicates a diagnosis test; it highlights that causes and consequences of a disease can be easily confused by non-specialists. Most of the left out sentences contain signs or symptoms expressed by complex units, such as Levels of creatinkinase in serum were high. (36%). 27% of these sentences are about gene mutations, which can be considered as causes of diseases or as clinical signs. Others contain patterns which have not been selected by the expert but can be easily added to improve the recall.

The context annotation is only a first step towards sign and symptom extraction. So far, we have not solved the problem of unit delimitation. In order to achieve this, we have two working hypotheses. We intend to compare chunking and syntactic analysis results in defining the scope of sign and symptom lexical units. Chunking will be conducted with an NLP tool such as TreeTagger, and syntactic analysis will use a dependency parser such as the Stanford Parser (ref.). The latter should allow us to delimit some recurring syntactic structures (e.g. agents, enumerations, etc.).

We also intend to compare our results with results provided by CRFs. First the features will be classical (bag of words, among others), and second, we will add the contexts obtained with the text mining to the features. This should enable us to compare our method to others. Finally, we are going to develop an evaluation interface to facilitate the work of the expert. In the absence of comparable corpora, the evaluation can only be manual. Our current sample of 50 abstracts is just a start, and needs to be expanded in order to strengthen the evaluation.

Acknowledgments

This research was supported by the Hybride Project ANR-11-BS02-002.

References

Rakesh Agrawal and Ramakrishnan Srikant. 1995. Mining Sequential Patterns. *Proceedings of ICDE'95.*

Daniel Albright, Arrick Lanfranchi, Anwen Fredriksen, William F. Styler IV, Colin Warner, Jena D. Hwang, Jinho D. Choi, Dmitriy Dligach, Rodney D. Nielsen, James Martin, Wayne Ward, Martha Palmer ans Guergana K. Savova. 2013. Towards comprehensive syntactic and semantic annotations of the clinical narrative. *Journal of the American Medical Informatics Association*, 20:922–930.

Nicolas Béchet, Peggy Cellier, Thierry Charnois and Bruno Crémilleux. 2012. Discovering linguistic patterns using sequence mining. *Proceedings of Springer LNCS, 13th International Conference on Intelligent Text Processing and Computational Linguistics - CICLing'2012*, 1:154–165.

Guozhu Dong and Jian Pei. 2007. *Sequence Data Mining.* Springer.

Carol Friedman. 1997. Towards a Comprehensive Medical Language Processing System: Methods and Issues. *Proceedings of the AMIA Annual Fall Symposium*, 1997:595–599.

Sebastian Köhler, Sandra C. Doelken, Christopher J. Mungall, Sebastian Bauer, Helen V. Firth, Isabelle Bailleul-Forestier, Graeme C. M. Black, Danielle L. Brown, Michael Brudno, Jennifer Campbell, David R. FitzPatrick, Janan T. Eppig, Andrew P. Jackson, Kathleen Freson, Marta Girdea, Ingo Helbig, Jane A. Hurst, Johanna Jähn, Laird G. Jackson, Anne M. Kelly, David H. Ledbetter, Sahar Mansour, Christa L. Martin, Celia Moss, Andrew Mumford, Willem H. Ouwehand, Soo-Mi Park, Erin Rooney Riggs, Richard H. Scott, Sanjay Sisodiya, Steven Van Vooren, Ronald J. Wapner, Andrew O. M. Wilkie, Caroline F. Wright, Anneke T. Vulto-van Silfhout, Nicole de Leeuw, Bert B. A. de Vries, Nicole L. Washingthon, Cynthia L. Smith, Monte Westerfield, Paul Schofield, Barbara J. Ruef, Georgios V. Gkoutos, Melissa Haendel, Damian Smedley, Suzanna E. Lewis and Peter N. Robinson. 2014. The Human Phenotype Ontology project: linking molecular biology and disease through phenotype data. *Nucleic Acids Research*, 42:966–974.

Dimitrios Kokkinakis. 2006. Developing Resources for Swedish Bio-Medical Text-Mining. *Proceedings of the 2nd International Symposium on Semantic Mining in Biomedicine (SMBM)*

Guergana K. Savova, James J. Masanz, Philip V. Ogren, Jiaping Zheng, Sunghwan Sohn, Karin C. Kipper-Schuler, Christopher G. Chute. 2010. Mayo clinical Text Analysis and Knowledge Extraction System (cTAKES): architecture, component evaluation and applications. *Journal of the American Medical Informatics Association*, 17:507–513.

Helmut Schmidt. 1994. Probabilistic Part-of-Speech Tagging Using Decision Trees. *Proceedings of International Conference on New Methods in Language Processing.*

Brett R. South, Shuying Shen, Makoto Jones, Jennifer Garvin, Matthew H. Samore, Wendy W. Chapman and Adi V. Gundlapalli. 2009. Developing a manually annotated clinical document corpus to identify phenotypic information for inflammatory bowel disease. *Summit on Translational Bioinformatics 2009*

Özlem Uzuner, Brett R. South, Shuying Shen, Scott L. DuVall. 2011. 2010 i2b2/VA challenge on concepts, assertions, and relations in clinical text. *Journal of the American Medical Informatics Association*, 18:552–556.

Kavishwar B. Wagholikar, Manabu Torii, Siddartha R. Jonnalagadda and Hongfang Liu. 2013. Pooling annotated corpora for clinical concept extraction. *Journal of Biomedical Semantics*, 4:3.

Alfred V. Aho and Jeffrey D. Ullman. 1972. *The Theory of Parsing, Translation and Compiling*, volume 1. Prentice-Hall, Englewood Cliffs, NJ.

Seeking Informativeness in Literature Based Discovery

Judita Preiss
University of Sheffield, Department of Computer Science
Regent Court, 211 Portobello
Sheffield S1 4DP, United Kingdom
j.preiss@sheffield.ac.uk

Abstract

The continuously increasing number of publications within the biomedical domain has fuelled the creation of literature based discovery (LBD) systems which identify unconnected pieces of knowledge appearing in separate literatures which can be combined to make new discoveries. Without filtering, the amount of hidden knowledge found is vast due to noise, making it impractical for a researcher to examine, or clinically evaluate, the potential discoveries. We present a number of filtering techniques, including two which exploit the LBD system itself rather than being based on a statistical or manual examination of document collections, and we demonstrate usefulness via replication of known discoveries.

1 Introduction and background

The number of publications in the biomedical domain has been observed to increase at a great rate, making it impossible for one person to read all, and thus potentially leaving knowledge hidden: for example, Swanson (1986) found one publication mentioning a connection between *Raynaud's Disease* and *blood viscosity* while another pointed out the effect of *fish oil* on *blood viscosity*, but there was no publication making the connection between *fish oil* and *Raynaud's Disease*. Automated approaches to knowledge discovery often set up the problem as outlined by Swanson; A being the source term (in this case *Raynaud's Disease*), with a possible target term, C, being specified (*fish oil*) and any connections between them form the linking, B, terms. If C is not specified, all possible hidden links from A are explored and discovery is classified as open. If both A and C terms are supplied, the discovery is closed and only any linking, B, terms are being sought.

Independent of how a connection between an A term and a B is defined (whether this is based on A and B co-occurring in the same title, in the same sentence or the same document, or some other relation), an obvious difficulty is the amount of data generated by a technique along these lines: with no filtering, a great number of connections will be made through terms such as *clinical study* or *patient*, and, if not also linked through other terms, these should be discarded. A number of approaches to term reduction have been explored.

Swanson and Smalheiser (1999)'s knowledge discovery system, Arrowsmith,[1] contains an increasing, currently 9,500 term (Swanson et al., 2006), stoplist, created semi-automatically.[2] Such a stoplist is unlikely to be complete – the list has grown from 5,000 (Swanson and Smalheiser, 1997) to 9,500 words (Swanson et al., 2006) and is likely to keep increasing. Over fitting is potentially an issue, in this case the list generated has been criticized for being tuned for the original *Raynaud–fish oil* discovery (Weeber et al., 2001). A word based stoplist also does not take into account the potential ambiguity of terms: one sense may be highly frequent and uninformative, guaranteeing it an appearance in the stoplist, while another sense may be rare but highly informative.

Instead of using words directly, it is possible to employ a (much smaller) controlled vocabulary: Medical Subject Headings (MeSH), consisting of 22,500 codes, are (mostly) manually assigned to each document indexed in Medline – even though multiple MeSH codes for a document are allowed, restricting to this set greatly reduces dimensionality. For example, Srinivasan (2004) uses MeSH based topic profiles to connect A to topics C via the most likely MeSH terms.

[1]Available at http://arrowsmith.psych.uic.edu/arrowsmith_uic/index.html

[2]Note that only 365 words of this stoplist are publicly available.

Proceedings of the 2014 Workshop on Biomedical Natural Language Processing (BioNLP 2014), pages 112–117,
Baltimore, Maryland USA, June 26-27 2014. ©2014 Association for Computational Linguistics

Keeping entire vocabularies is possible if topics are limited, for example, Fleuren et al (2011) extract statistics regarding gene co-occurrence, and restricts their hidden knowledge generation to biological mechanisms related to them.

Another difficulty in using word vocabularies is the necessary identification of multiwords, Weeber at al. (2001) avoid previously tried n-gram techniques (e.g. (Gordon and Lindsay, 1996)) by switching knowledge discovery to UMLS Concept Unique Identifiers (CUIs). Using MetaMap (Rindflesch and Aronson, 1994) to assign CUIs to texts discards non content words (CUIs only exist for concepts), resolves ambiguity and deals with multiwords in one, thus reducing the number of terms considered in later stages. Weeber et al. also exploit the broad subject categories that UMLS assigns to each CUI, which allow the authors to perform domain specific filtering to reduce dimensionality. This they do on a per search basis, tuning the filtering to the replication experiments presented.

Dimensionality reduction can also be performed at the relation level. Swanson's (1997) original work deemed two terms connected if they both appeared in the title of an abstract – titles were thought to be the most informative, and descriptive, part of each article. As the number of abstracts explored during the knowledge discovery process increased, and connections were extended to whole abstracts (rather than titles only), the amount of hidden knowledge generated increased dramatically and with it did the need for term and connection filtering.

Hristovski et al (2006) argue for filtering within the relation definition – co-occurrence does not provide any basis for a relation between two terms, no underlying semantic reason, and thus, as well as leading to many spurious links, it yields no justification for a hidden connection that is found. They extract subject-relation-object triples, with relations such as *treats* or *affects* forming their UMLS concept relations, leading to a much smaller number of (more accurate) relations to derive hidden knowledge from.

While re-ranking (placing the most 'useful' links at the top of the list) the resulting hidden knowledge is clearly valuable, removing terms from consideration prior to identifying hidden knowledge will reduce the computational load as well as avoid noisy hidden knowledge being produced and possibly accidentally being highly ranked.

We explore a number of filtering approaches including two novel techniques which can be integrated into any method designed using the Swanson framework, and we compare these against previously explored filtering methods. Section 2 outlines our knowledge discovery approach, Section 3 presents a number of filtering approaches with Section 4 discussing results based on replication of existing knowledge and Section 5 draws our conclusions.

2 Knowledge discovery system

There are two main components which define an LBD system created following the Swanson framework: the terms and the relations. Based on arguments presented in Section 1, our system employs UMLS CUIs as produced by SemRep (Rindflesch and Fiszman, 2003), a natural language processing system which identifies semantic relations in biomedical text.[3]

SemRep extracts relation triples from text by running a set of rules over the output of an underspecified parser. The rules, such as the mapping of *treatment* to TREATS, map syntactic indicators to predicates in the Semantic Network. Further restrictions are imposed regarding the permissibility of arguments, the viability of the given propositions, and other syntactic constraints, resulting in relations such as

- Epoprostenol TREATS Raynaud Phenomenon
- blood rheology DIAGNOSES Raynaud Disease

Each triple is also output with the corresponding CUIs.

All 29 non negative relations were extracted (such as AFFECTS, ASSOCIATED_WITH, INTERACTS_WITH, ...), while negative relations (such as NEG_AFFECTS, NEG_ASSOCIATED_WITH, NEG_INTERACTS_WITH, ...) were dropped. The extracted relations form the connections between CUIs: i.e., the set of linking CUIs B is created by following all SemRep links from the CUI A, which lead to C through another SemRep relation.

[3]In this work, the SemRep annotated Medline data, database semmedVER24 (processed up to November 2013) run over 23,319,737 citations to yield 68,000,470 predications, was downloaded from `http://skr3.nlm.nih.gov` and used throughout.

3 Filtering approaches

While employing CUIs (rather than words) eliminates non content words (thus immediately reducing noise), it does not eliminate CUIs corresponding to *patient*, *week*, *statement* ... We present, and in Section 4 evaluate (individually and in combination), four filtering approaches of which two are, to our knowledge, completely novel.

3.1 Synonyms

While not a filtering method under the usual definition, the identification of synonym CUIs and collapsing thereof results in the reduction of the number of CUIs being used (i.e. the technique filters out some CUIs).

A manual examination of the documents containing CUI C0034734, *Raynaud Disease*, revealed that some of the expected connections were missing and were linked to CUI C0034735, *Raynaud Phenomenon*, instead. The resulting hidden knowledge is greatly affected by the particular CUI chosen as the source term A, yet in this case, the two CUIs are synonymous. The MRREL related concepts file within UMLS contains pairs of CUIs within related relationships, including the SY (source asserted synonymy) relationship[4], and CUIs C0034734 and C0034735 appear in the SY relationship in this list. Identifying concepts within the SY relationship has the following advantages:

- Merging such synonyms into classes will allow the retrieval of more hidden knowledge if the multiple synonymous CUIs correspond to the start point, A (as in the case of *Raynaud Disease*).
- There will be potentially more hidden knowledge created if a multiclass CUI is a linking term (as A connected to C0034734 and C connected to C0034735 would not have been found to be connected if these were the only potential overlap).
- Synonymous hidden knowledge (and linking terms) will merge, reducing the amount of knowledge (and terms) to manually explore.

Merging synonyms into single CUI classes reduces the 561,155 CUIs present in UMLS to 540,440 CUI classes.[5]

3.2 Semantic types

The UMLS Semantic Network consists of 133 semantic types, a type of subject category, which is assigned to each CUI. Many of these categories are clearly unhelpful for knowledge discovery (for example, *geographic area* or *language*), and 70 semantic types are manually selected for removal (by examining the basic information about the relation, as well as the structure of the network and the CUIs assigned each semantic type). This removes a further 121,284 CUIs.

3.3 Discarding common linking terms

In some cases, a given CUI is clearly too general to be a useful linking term, but its semantic type contains more specific CUIs which should not be removed. Restricting semantic type filtering based on the depth within the hierarchy is also not a viable option, as UMLS is composed of different hierarchies, each with a different level of granularity and establishing an overall threshold would likely include general terms for some while discarding crucial terms for others. Therefore another approach is needed for these CUIs.

Along the lines of Swanson et al (2006), a stoplist can be built to contain such terms, without over-training for a particular discovery and without the need for manual intervention: we hypothesize that any CUIs which are linking terms more often than others can effectively form a stoplist.

The creation of this stoplist can be performed iteratively:

1. Start with an empty stoplist set S.
2. Create hidden knowledge based on SemRep connections between CUIs, removing any connections to CUIs in set S (the hidden knowledge is acquired from Medline articles published between 1865 and 2000).
3. Randomly select 10,000 hidden knowledge pairs, identify their linking CUIs, and add any linking CUIs appearing in more than *threshold* of pairs to S (the value of *threshold* needs to be empirically determined).
4. If Step 3 increased the size of S, return to Step 2.

Note that since the training set is not designed for any particular discovery, this should not result in an over trained stoplist.

[4] Due to the version of SemRep files used, UMLS 2013AA is employed throughout.

[5] Note that other MRREL related relationships were explored, but completing cycles lead to multiple extremely large equivalence classes.

3.4 Breaking high frequency connections

The creation of a stoplist will always suffer from omissions and inclusions of CUIs that should not be filtered out in every instance. The last approach is based on a slightly different underlying idea: instead of finding frequently appearing terms, this approach bases its decisions on the number of terms a given term is connected to.

Two CUIs A and B are deemed connected if a (non negative) SemRep relation exists which links them. If A corresponds to a term such as *study* or *patient*, it is expected to be connected to a large number of CUIs. We hypothesize that terms which are so highly connected are likely to be relatively general terms, and so uninformative linking terms.

This gives rise to the following filtering options:

1. Break (discard) all connections to CUI A when the $C(A) >$ threshold.

2. Discard the connection between CUIs A and B when $min(C(A), C(B)) >$ threshold.

(Where $C(A)$ represents the number of CUIs linked to A, and the threshold needs to be empirically determined.)

Method 1 effectively forms a stoplist of highly connected CUIs, but method 2 is different: only connections satisfying the condition are broken while A remains under consideration. This allows filtering method 2 to leave a frequently connected term to be a linking term for a rare term (unlike method 1, which would discard such a term).

4 Results

Swanson's original discoveries (Swanson, 1986; Swanson, 1988) were verified through clinical trials and evaluation of LBD systems often involves replication of these discoveries (Gordon and Lindsay, 1996; Weeber et al., 2001). From literature, we identify seven separate discoveries to replicate (presented with the labels used in Table 1):

RD: Raynaud disease and fish oil (Swanson, 1986).

Arg: Somatomedin C and arginine (Swanson, 1990).

Mg: Migraine disorders and magnesium (Hu et al., 2006).

ND: Magnesium deficiency and neurologic disease (Smalheiser and Swanson, 1994).

INN: Alzheimer's and indomethacin (Smalheiser and Swanson, 1996a).

estrogen: Alzheimer's disease and estrogen (Smalheiser and Swanson, 1996b).

Ca^{2+}iPLA2: Schizophrenia and Calcium-Independent Phospholipase A2 (Smalheiser and Swanson, 1997).

The same subset of Medline as in each original discovery is employed for replication, and any abstracts containing a direct link between the two terms are removed (note that including these would not have affected the original discoveries as these only used titles) – thus any connections between A and C are necessarily hidden and require at least one linking term.

The Raynaud-fish oil and migraine-magnesium connections are the most commonly replicated discoveries, while the remaining discoveries are rarely explored. For CUI based investigations, this is likely due to the difficulty of selecting a representative CUI for the sought concepts. The second concept in the Schizophrenia and Calcium-Independent Phospholipase A2 connection is particularly tricky: UMLS suggests CUI C1418624 (PLA2G6 gene) as the most likely match, followed by CUI C2830173 (Calcium-Independent Phospholipase A2) as the second most likely. However, neither CUI is found in any relations in the given date range by SemRep. Closer examination reveals that the Ca^{2+}iPLA2 connections in the 1960-1997 Medline range are between CUI C0538273 (PLA2G6 protein, human). Not only does this highlight the difficulty of the replication task, it further motivates the need for a 'synonym' (or related concept) list.

The number of linking terms found between each pair of sought terms is presented in Table 1 (zero linking terms means the connection was not found) for a subset of the filtering results. ST represents semantic type filtering, HF the breaking of high frequency connections (a *min* subscript denoting the version which takes into account connectivity of both CUIs), together with the threshold value, and LT elimination of common linking terms, again with the relevant threshold value.

While the Raynaud-fish oil connection appears to be consistently produced by the system, Table 2 reveals the value of filtering: with no filtering, the two linking terms are pure noise and the connection should not be made. Employing UMLS syn-

	RD	Arg	Mg	ND	INN	estrogen	Ca^{2+}iPLA2
No filtering	2	235	78	98	370	500	7
Synonyms (Sy)	6	173	58	65	296	415	16
Sy & LT-200	3	145	48	56	265	0	16
Sy & HF-2900	6	149	56	52	243	0	14
Sy & HF_{min}-900	6	73	22	27	82	164	9
Sy & HF_{min}-400	6	25	5	8	25	65	8
Sy & ST	4	130	47	43	234	331	13
Sy & ST & LT-200	3	108	41	38	207	0	13
Sy & ST & HF-2500	4	120	47	38	205	0	13
Sy & ST & HF_{min}-900	6	73	22	27	82	164	9
Sy & ST & HF_{min}-400	4	28	6	12	30	73	6

Table 1: Number of hidden links found during replication

onyms adds genuine linking terms,[6] and restricting by semantic types drops the remaining general terms. Discarding common linking terms finds *antimicrobial susceptibility* to be a frequently used linking term, and it is also dropped. A great advantage of the technique can be seen when connections are made through hundreds of terms – in this case, higher thresholds (and thus more aggressive filtering) can be employed to reduce the number of linking terms to the most promising set. Should these not be sufficient, the threshold can be increased to produce more linking terms and as such, the burden on the user in checking a large number of linking terms when a hidden connection is suspected can be greatly reduced, without sacrificing connections should more be needed.

Term	NF	Sy	Sy ST	LT-200
acetylsalicyclic acid	×	✓	✓	✓
antimicrobial susceptibility	×	✓	✓	×
blood viscosity	×	✓	✓	✓
brain infarction	×	✓	✓	✓
patient	✓	✓	×	×
volunteer helper	✓	✓	×	×

Table 2: Linking term analysis for RD

For example, common linking term filtering removes the term *estrogen* from consideration as *therapeutic estrogen* is a commonly used linking term, making the *estrogen-AD* link impossible to find. Linking term frequencies (on a 10,000 pair sample) exceeding values from 50 to 200 (in increments of 50) were tested resulting in the removal of between 1,902 and 227 CUIs. *Therapeutic estrogen* appears in all the lists. Similarly, the CUI is dropped when high frequency connections are broken using the first technique, which is based on stoplists. This highlights the value of the second high frequency connection technique, which only discards particular connections (rather than CUIs) and *therapeutic estrogen* CUI remains a searchable CUI.

As shown, the system replicates most of the previously published discoveries with its main asset being noise reduction: the number of linking terms for a suspected connection (closed discovery) can be greatly reduced to remove spurious connections, with backoffs available to yield more connections should more be required. For novel applications (i.e. open discovery), the technique greatly reduces the amount of hidden knowledge generated from a source term A. For example, the amount of hidden knowledge generated from somatomedin C drops from 82,601 CUIs when no filtering is performed, to 3,005 CUIs with synonym, semantic type and breaking connections with frequency more than 200, which represents a great reduction for a user who is likely looking for a particular type of C term.

5 Conclusions and future work

We present and demonstrate the effectiveness of a number of filtering methods, including two novel techniques based on any LBD system built according to the Swanson framework – one approach based on stoplist methods, but requiring no manual intervention except for a user's selection of a threshold, and the second based on removing connections when these are deemed to be likely to

[6] Note that the merging of synonyms is achieved without the need to back off to general classes (e.g. (Srinivasan, 2004)), which have been observed to lead to connections based on "aboutness" rather than producing genuine hidden knowledge (Beresi et al., 2008).

contribute mainly noise. A great advantage of the second approach is shown to be the fact that terms are not directly discarded, as with a stoplist, and thus a fairly common term can remain a source term when required.

While the method is evaluated by replicating known discoveries, we suggest that the noise reduction performed is ultimately leading to a much more user friendly LBD system, and plan to investigate other evaluation approaches, such as timeslicing (Yetisgen-Yildiz and Pratt, 2009), as part of future work.

Acknowledgements

Judita Preiss was supported by the EPSRC grant EP/J008427/1: Language Processing for Literature Based Discovery in Medicine.

References

Ulises Cervino Beresi, Mark Baillie, and Ian Ruthven. 2008. Towards the evaluation of literature based discovery. In *Proceedings of the Workshop on Novel Evaluation Methodologies (at ECIR 2008)*, pages 5–13.

Wilco W. M. Fleuren, Stefan Verhoeven, Raoul Frijters, Bart Heupers, Jan Polman, René van Schaik, Jacob de Vlieg, and Wynand Alkema. 2011. Copub update: Copub 5.0 a text mining system to answer biological questions. *Nuclear Acids Research*, 39 (Web Server issue). doi:10.1093/nar/gkr310.

Michael D. Gordon and Robert K. Lindsay. 1996. Toward discovery support systems: a replication, reexamination, and extension of Swanson's work on literature-based discovery of a connection between Raynaud's and fish oil. *Journal of the American Society for Information Science*, 47(2):116–128.

Thomas C. Rindflesch Hristovski D, Friedman C and Peterlin B. 2006. Exploiting semantic relations for literature-based discovery. In *Proceedings of the 2006 AMIA Annual Symposium*, pages 349–353.

Xiaohua Hu, Xiaodan Zhang, Illhoi Yoo, and Yanquing Zang. 2006. A semantic approach for mining hidden links from complementary and non-interactive biomedical literature. In *SDM*.

Thomas C. Rindflesch and Alan R. Aronson. 1994. Ambiguity resolution while mapping free text to the UMLS Metathesaurus. In J. G. Ozbolt, editor, *Proceedings of the Eigheeth Annual Symposium on Computer Appplications in Medical Care*, pages 240–244.

Thomas C. Rindflesch and Marcelo Fiszman. 2003. The interaction of domain knowledge and linguistic structure in natural language processing: interpreting hypernymic propositions in biomedical text. *Journal of Biomedical Informatics*, 36(6):462–477.

Neil R. Smalheiser and Don R. Swanson. 1994. Assessing a gap in the biomedical literature: Magnesium deficiency and neurologic disease. *Neuroscience Research Communications*, 15(1):1–9.

Neil R. Smalheiser and Don R. Swanson. 1996a. Indomethacin and Alzheimer's disease. *Neurology*, 46:583.

Neil R. Smalheiser and Don R. Swanson. 1996b. Linking estrogen to Alzheimer's disease. *Neurology*, 47:809–810.

Neil R. Smalheiser and Don R. Swanson. 1997. Calcium-independent phospholipase a2 and schizophrenia. *Arch Gen Psychiatry*, 55(8):752–753.

Padmini Srinivasan. 2004. Text mining generating hypotheses from medline. *Journal of the American Society for Information Science and Technology*, 55(5):396–413.

Don R. Swanson and Neil R. Smalheiser. 1997. An interactive system for finding complementary literatures: A stimulus to scientific discovery. *Artificial Intelligence*, 91:183–203.

Don R. Swanson and Neil R. Smalheiser. 1999. Link analysis of MEDLINE titles as an aid to scientific discovery: Using Arrowsmith as an aid to scientific discovery. *Library Trends*, 48:48–59.

Don R. Swanson, Neil R. Smalheiser, and Vetle I. Torvik. 2006. Ranking indirect connnections in literature-based discovery: The role of medical subject headings. *Journal of the American Society for Information Science and Technology*, 57(11):1427–1439.

Don R. Swanson. 1986. Fish oil, Raynaud's syndrome, and undiscovered public knowledge. *Perspectives in Biology and Medicine*, 30:7–18.

Don R. Swanson. 1988. Migraine and magnesium – 11 neglected connections. *Perpectives in Biology and Medicine*, 31(4):526–557.

Don R. Swanson. 1990. Somatomedin c and arginine: Implicit connections between mutually isolated literatures. *Perspectives in Biology and Medicine*, 33(2):157–186.

Marc Weeber, Rein Vos, Henny Klein, and Lolkje T. W. de Jong-van den Berg. 2001. Using concepts in literature-based discovery: Simulating Swanson's Reynaud – fish oil and migraine – magnesium discoveries. *Journal of the American Society for Information Science and Technology*, 52(7):548–557.

M. Yetisgen-Yildiz and W. Pratt. 2009. A new evaluation methodology for literature-based discovery. *Journal of Biomedical Informatics*, 42(4):633–643.

Towards Gene Recognition from Rare and Ambiguous Abbreviations using a Filtering Approach

Matthias Hartung*, **Roman Klinger***, **Matthias Zwick**‡ **and Philipp Cimiano***

*Semantic Computing Group
Cognitive Interaction Technology –
Center of Excellence (CIT-EC)
Bielefeld University
33615 Bielefeld, Germany
{mhartung,rklinger,cimiano}
@cit-ec.uni-bielefeld.de

‡Research Networking
Boehringer Ingelheim Pharma GmbH
Birkendorfer Str. 65
88397 Biberach, Germany
matthias.zwick
@boehringer-ingelheim.com

Abstract

Retrieving information about highly ambiguous gene/protein homonyms is a challenge, in particular where their non-protein meanings are more frequent than their protein meaning (*e. g.*, *SAH* or *HF*). Due to their limited coverage in common benchmarking data sets, the performance of existing gene/protein recognition tools on these problematic cases is hard to assess.

We uniformly sample a corpus of eight ambiguous gene/protein abbreviations from MEDLINE® and provide manual annotations for each mention of these abbreviations.[1] Based on this resource, we show that available gene recognition tools such as conditional random fields (CRF) trained on BioCreative 2 NER data or GNAT tend to underperform on this phenomenon.

We propose to extend existing gene recognition approaches by combining a CRF and a support vector machine. In a cross-entity evaluation and without taking any entity-specific information into account, our model achieves a gain of 6 points F_1-Measure over our best baseline which checks for the occurrence of a long form of the abbreviation and more than 9 points over all existing tools investigated.

1 Introduction

In pharmaceutical research, a common task is to gather all relevant information about a gene, *e. g.*, from published articles or abstracts. The task of recognizing the mentions of genes or proteins can be understood as the classification problem to decide whether the entity of interest denotes a gene/protein or something else. For highly ambiguous short names, this task can be particularly challenging. Consider, for instance, the gene *acyl-CoA synthetase medium-chain family member 3* which has synonyms *protein SA homolog* or *SA hypertension-associated homolog*, among others, with abbreviations *ACSM3*, and *SAH*.[2] Standard thesaurus-based search engines would retrieve results where *SAH* denotes the gene/protein of interest, but also occurrences in which it denotes other proteins (*e. g.*, *ATX1 antioxidant protein 1 homolog*[3]) or entities from semantic classes other than genes/proteins (*e. g.*, the symptom *sub-arachnoid hemorrhage*).

For an abbreviation such as *SAH*, the use as denoting a symptom or another semantic class different from genes/proteins is more frequent by a factor of 70 compared to protein-denoting mentions according to our corpus analysis, such that the retrieval precision for *acyl-CoA synthetase* by the occurrence of the synonym *SAH* is only about 0.01, which is totally unacceptable for practical applications.

In this paper, we discuss the specific challenge of recognizing such highly ambiguous abbreviations. We consider eight entities and show that common corpora for gene/protein recognition are of limited value for their investigation. The abbreviations we consider are SAH, MOX, PLS, CLU, CLI, HF, AHR and COPD (*cf.* Table 1). Based on a sample from MEDLINE[4], we show that these names do actually occur in biomedical text, but are underrepresented in corpora typically used for benchmarking and developing gene/protein recognition approaches.

[1] The annotated corpus is available for future research at `http://dx.doi.org/10.4119/unibi/2673424`.

[2] `http://www.ncbi.nlm.nih.gov/gene/6296`

[3] `http://www.ncbi.nlm.nih.gov/gene/443451`

[4] `http://www.nlm.nih.gov/pubs/factsheets/medline.html`

Proceedings of the 2014 Workshop on Biomedical Natural Language Processing (BioNLP 2014), pages 118–127,
Baltimore, Maryland USA, June 26-27 2014.

Synonym	Other names	Other meaning	EntrezGene ID
SAH	acyl-CoA synthetase medium-chain family member 3; ACSM3	subarachnoid hemorrhage; S-Adenosyl-L-homocysteine hydrolase	6296
MOX	monooxygenase, DBH-like 1	moxifloxacin; methylparaoxon	26002
PLS	POLARIS	partial least squares; primary lateral sclerosis	3770598
CLU	clusterin; CLI	covalent linkage unit	1191
CLI	clusterin; CLU	clindamycin	1191
HF	complement factor H; CFH	high frequency; heart failure; Hartree-Fock	3075
AHR	aryl hydrocarbon receptor; bHLHe76	airway hyperreactivity	196
COPD	archain 1; ARCN1; coatomer protein complex, subunit delta	Chronic Obstructive Pulmonary Disease	22819; 372

Table 1: The eight synonyms for genes/proteins which are subject of analysis in this paper and their long names together with frequent other meanings.

We propose a machine learning-based filtering approach to detect whether a mention in question actually denotes a gene/protein or not and show that for the eight highly ambiguous abbreviations that we consider, the performance of our approach in terms of F_1 measure is higher than for a state-of-the-art tagger based on conditional random fields (CRF), a freely available dictionary-based approach and an abbreviation resolver. We evaluate different parameters and their impact in our filtering approach and discuss the results. Note that this approach does not take any information about the specific abbreviation into account and can therefore be expected to generalize to names not considered in our corpus.

The main contributions of this paper are:

(i) We consider the problem of recognizing highly ambiguous abbreviations that frequently do not denote proteins as a task that has so far attracted only limited attention.

(ii) We show that the recognition of such ambiguous mentions is important as their string representation is frequent in collections such as MEDLINE.

(iii) We show, however, that this set of ambiguous names is underrepresented in corpora commonly used for system design and development. Such corpora do not provide a sufficient data basis for studying the phenomenon or for training systems that appropriately handle such ambiguous abbreviation. We contribute a manually annotated corpus of 2174 occurrences of ambiguous abbreviations.

(iv) We propose a filtering method for classifying ambiguous abbreviations as denoting a protein or not. We show that this method has a positive impact on the overall performance of named entity recognition systems.

2 Related Work

The task of gene/protein recognition consists in the classification of terms as actually denoting a gene/protein or not. The task is typically either tackled by using machine learning or dictionary-based approaches. Machine learning approaches rely on appropriate features describing the local context of the term to be classified and induce a model to perform the classification from training data. Conditional random fields have shown to yield very good results on the task (Klinger et al., 2007; Leaman and Gonzalez, 2008; Kuo et al., 2007; Settles, 2005).

Dictionary-based approaches rely on an explicit dictionary of gene/protein names that are matched in text. Such systems are common in practice due to the low overhead required to adapt and maintain the system, essentially only requiring to extend the dictionary. Examples of commercial systems are ProMiner (Fluck et al., 2007) or I2E (Bandy et al., 2009); a popular free system is made available by Hakenberg et al. (2011).

Such dictionary-based systems typically incorporate rules for filtering false positives. For instance, in ProMiner (Hanisch et al., 2003), ambiguous synonyms are only accepted based on external dictionaries and matches in the context. Abbreviations are only accepted if a long form matches all parts of the abbreviation in the context (following Schwartz and Hearst (2003)). Similarly, Hakenberg et al. (2008) discuss global disambiguation on the document level, such that all mentions of a string in one abstract are uniformly accepted as denoting an entity or not.

A slightly different approach is taken by the web-service GeneE[5] (Schuemie et al., 2010): Entering a query as a gene/protein in the search field generates

[5]http://biosemantics.org/geneE

Protein	MEDLINE #Tokens	MEDLINE % tagged	BioCreative2 #Tokens	BioCreative2 % of genes	GENIA #Tokens	GENIA % of genes
SAH	30019	6.1 %	2	0 %	0	
MOX	16007	13.1 %	0		0	
PLS	11918	25.9 %	0		0	
CLU	1077	29.1 %	0		0	
CLI	1957	4.8 %	4	0 %	0	
HF	42563	7.9 %	8	62.5 %	4	0 %
AHR	21525	75.7 %	12	91.7 %	0	
COPD	44125	0.6 %	6	0 %	0	

Table 2: Coverage of ambiguous abbreviations in MEDLINE, BioCreative2 and GENIA corpora. The percentage of tokens tagged as a gene/protein in MEDLINE (% tagged) is determined with a conditional random field in the configuration described by Klinger et al. (2007), but without dictionary-based features to foster the usage of contextual features). The percentages of genes/proteins (% of genes) in BC2 and GENIA are based on the annotations in these corpora.

a query to *e. g.* PubMed®[6] with the goal to limit the number of false positives.

Previous to the common application of CRFs, other machine learning methods have been popular as well for the task of entity recognition. For instance, Mitsumori et al. (2005) and Bickel et al. (2004) use a support vector machine (SVM) with part-of-speech information and dictionary-based features, amongst others. Zhou et al. (2005) use an ensemble of different classifiers for recognition.

In contrast to this application of a classifier to solve the recognition task entirely, other approaches (including the one in this paper) aim at filtering specifically ambiguous entities from a previously defined set of challenging terms. For instance, Al-mubaid (2006) utilize a word-based classifier and a mutual information-based feature selection to achieve a highly discriminating list of terms which is applied for filtering candidates.

Similarly to our approach, Tsuruoka and Tsujii (2003) use a classifier, in their case a naïve Bayes approach, to learn which entities to filter from the candidates generated by a dictionary-based approach. They use word based features in the context including the candidate itself. Therefore, the approach is focused on specific entities.

Gaudan et al. (2005) use an SVM and a dictionary of long forms of abbreviations to assign them a specific meaning, taking contextual information into account. However, their machine learning approach is trained on each possible sense of an abbreviation. In contrast, our approach consists in deciding if a term is used as a protein or not. Further, we do not train to detect specific, previously given senses.

[6]http://www.ncbi.nlm.nih.gov/pubmed/

Xu et al. (2007) apply text similarity measures to decide about specific meanings of mentions. They focus on the disambiguation between different entities. A corpus for word sense disambiguation is automatically built based on MeSH annotations by Jimeno-Yepes et al. (2011). Okazaki et al. (2010) build a sense inventory by automatically applying patterns on MEDLINE and use this in a logistic regression approach.

Approaches are typically evaluated on freely available resources like the BioCreative Gene Mention Task Corpus, to which we refer as BC2 (Smith et al., 2008), or the GENIA Corpus (Kim et al., 2003). When it comes to identifying particular proteins by linking the protein in question to some protein in an external database – a task we do not address in this paper – the BioCreative Gene Normalization Task Corpus is a common resource (Morgan et al., 2008).

In contrast to these previous approaches, our method is not tailored to a particular set of entities or meanings, as the training methodology abstracts from specific entities. The model, in fact, knows nothing about the abbreviations to be classified and does not use their surface form as a feature, such that it can be applied to any unseen gene/protein term. This leads to a simpler model that is applicable to a wide range of gene/protein term candidates. Our cross-entity evaluation regime clearly corroborates this.

3 Data

We focus on eight ambiguous abbreviations of gene/protein names. As shown in Table 2, these homonyms occur relatively frequently in MEDLINE but are underrepresented in the BioCreative 2 entity

Protein	Pos. Inst.	Neg. Inst.	Total
SAH	5	349	354
MOX	62	221	283
PLS	1	206	207
CLU	235	30	265
CLI	11	211	222
HF	2	353	355
AHR	53	80	133
COPD	0	250	250

Table 3: Number of instances per protein in the annotated data set and their positive/negative distribution

recognition data set and the GENIA corpus which are both commonly used for developing and evaluating gene recognition approaches. We compiled a corpus from MEDLINE by randomly sampling 100 abstracts for each of the eight abbreviations (81 for MOX) such that each abstract contains at least one mention of the respective abbreviation. One of the authors manually annotated the mentions of the eight abbreviations under consideration to be a gene/protein entity or not. These annotations were validated by another author. Both annotators disagreed in only 2% of the cases. The numbers of annotations, including their distribution over positive and negative instances, are summarized in Table 3. The corpus is made publicly available at `http://dx.doi.org/10.4119/unibi/2673424` (Hartung and Zwick, 2014).

In order to alleviate the imbalance of positive and negative examples in the data, additional positive examples have been gathered by manually searching PubMed[7]. At this point, special attention has been paid to extract only instances denoting the correct gene/protein corresponding to the full long name, as we are interested in assessing the impact of examples of a particularly high quality. This process yields 69 additional instances for AHR (distributed over 11 abstracts), 7 instances (3 abstracts) for HF, 14 instances (2 abstracts) for PLS and 15 instances (7 abstracts) for SAH. For the other gene/proteins in our dataset, no additional positive instances of this kind could be retrieved using PubMed. In the following, this process will be referred to as *manual instance generation*. This additional data is used for training only.

[7] `http://www.ncbi.nlm.nih.gov/pubmed`

4 Gene Recognition by Filtering

We frame gene/protein recognition from ambiguous abbreviations as a filtering task in which a set of candidate tokens is classified into entities and non-entities. In this paper, we assume the candidates to be generated by a simple dictionary-based approach taking into account all tokens that match the abbreviation under consideration.

4.1 Filtering Strategies

We consider the following filtering approaches:

- *SVM* classifies the occurring terms based on a binary support vector machine.
- *CRF* classifies the occurring terms based on a conditional random field (configured as described by Klinger et al. (2007)) trained on the concatenation of BC2 data and our newly generated corpus. This setting thus corresponds to state-of-the-art performance on the task.
- *CRF∩SVM* considers the candidate an entity if both the standard CRF and the SVM from the previous steps yield a positive prediction.
- *HRCRF∩SVM* is the same as the previous step, but the output of the CRF is optimized towards high recall by joining the recognition of entities of the five most likely Viterbi paths.
- *CRF→SVM* is similar to the first setting, but the output of the CRF is taken into account as a feature in the SVM.

4.2 Features for Classification

Our classifier uses local contextual and global features. Local features focus on the immediate context of an instance, whereas global features encode abstract-level information. Throughout the following discussion, t_i denotes a token at position i that corresponds to a particular abbreviation to be classified in an abstract A. Note that we blind the actual representation of the entity to be able to generalize to all genes/proteins, not being limited to the ones contained in our corpus.

4.2.1 Local Information

The feature templates *context-left* and *context-right* collect the tokens immediately surrounding an abbreviation in a window of size 6 (left) and 4 (right) in a bag-of-words-like feature generation. Additionally, the two tokens from the immediate context on each side are combined into bigrams.

The template *abbreviation* generates features if t_i occurs in brackets. It takes into account the minimal Levenshtein distance (*ld*, Levenshtein (1966)) between all long forms L of the abbreviation (as retrieved from EntrezGene) in comparison to each string on the left of t_i (up to a length of seven, denoted by $t_{k:i}$ as the concatenation of tokens $t_k, \ldots, t_i$). Therefore, the similarity value $sim(t_i)$ taken into account is given by

$$sim(t_i) = \max_{l \in L; k \in [1:7]} 1 - \frac{ld(t_{k:i-1}, l)}{\max(|t_i|, |l|)},$$

where the denominator is a normalization term. The features used are generated by cumulative binning of $sim(t_i)$.

The feature $tagger_{local}$ takes the prediction of the CRF for t_i into account. Note that this feature is only used in the CRF→SVM setting.

4.2.2 Global Information

The feature template *unigrams* considers each word in A as a feature. There is no normalization or frequency weighting. Stopwords are ignored[8]. Occurrences of the same string as t_i are blinded.

The feature $tagger_{global}$ collects all tokens in A other than t_i that are tagged as an entity by the CRF. In addition, the cardinality of these entities in A is taken into account by cumulative binning.

The feature *long form* holds if one of the long forms previously defined to correspond with the abbreviation occurs in the text (in arbitrary position).

Besides using all features, we perform a greedy search for the best feature set by wrapping the best model configuration. A detailed discussion of the feature selection process follows in Section 5.3.

4.2.3 Feature Propagation

Inspired by the "one sense per discourse" heuristic commonly adopted in word sense disambiguation (Gale et al., 1992), we apply two feature combination strategies. In the following, n denotes the number of occurrences of the abbreviation in an abstract.

In the setting $propagation_{all}$, $n - 1$ identical *linked instances* are added for each occurrence. Each new instance consists of the disjunction of the feature vectors of all occurrences. Based on the intuition that the first mention of an abbreviation might carry particularly valuable information, $propagation_{first}$ introduces one additional linked instance for each occurrence, in which the feature vector is joined with the first occurrence.

[8]Using the stopword list at `http://www.ncbi.nlm.nih.gov/books/NBK3827/table/pubmedhelp.T43/`, last accessed on March 25, 2014

Setting	P	R	F_1
SVM	0.81	0.45	0.58
CRF∩SVM	**0.99**	0.26	0.41
HRCRF∩SVM	0.95	0.27	0.42
CRF→SVM	0.83	0.49	0.62
CRF→SVM+FS	0.97	**0.74**	**0.84**
GNAT	0.73	0.45	0.56
CRF	0.55	0.43	0.48
AcroTagger	0.92	0.63	0.75
Long form	0.98	0.65	0.78
lex	0.18	1.00	0.32

Table 4: Overall micro-averaged results over eight genes/proteins. For comparison, we show the results of a default run of GNAT (Hakenberg et al., 2011), a CRF trained on BC2 data (Klinger et al., 2007), AcroTagger (Gaudan et al., 2005), and a simple approach of accepting every token of the respective string as a gene/protein entity (lex). Feature selection is denoted with +FS.

In both settings, all original and linked instances are used for training, while during testing, original instances are classified by majority voting on their linked instances. For $propagation_{all}$, this results in classifying each occurrence identically.

5 Experimental Evaluation

5.1 Experimental Setting

We perform a cross-entity evaluation, in which we train the support vector machine (SVM) on the abstracts of 7 genes/proteins from our corpus and test on the abstracts for the remaining entities, *i. e.*, the model is evaluated only on tokens representing entities which have never been seen labeled during training. The CRFs are trained analogously with the difference that the respective set used for training is augmented with the BioCreative 2 Training data. The average numbers of precision, recall and F_1 measure are reported.

As a baseline, we report the results of a simple lexicon-based approach assuming that all tokens denote an entity in all their occurrences (lex). In addition, the baseline of accepting an abbreviation as gene/protein if the long form occurs in the same abstract is reported (Long form). Moreover, we compare our results with the publicly available toolkit GNAT (Hakenberg et al., 2011)[9] and the CRF ap-

[9]The gene normalization functionality of GNAT is not taken into account here. We acknowledge that this comparison

proach as described in Section 4. In addition, we take into account the AcroTagger[10] that resolves abbreviations to their most likely long form which we manually map to denoting a gene/protein or not.

5.2 Results

5.2.1 Overall results

In Table 4, we summarize the results of the recognition strategies introduced in Section 4. The lexical baseline clearly proves that a simple approach without any filtering is not practical. GNAT adapts well to ambiguous short names and turns out as a competitive baseline, achieving an average precision of 0.73. In contrast, the filtering capacity of a standard CRF is, at best, mediocre. The long form baseline is very competitive with an F_1 measure of 0.78 and a close-to-perfect precision. The results of AcroTagger are similar to this long form baseline.

We observe that the SVM outperforms the CRF in terms of precision and recall (by 10 percentage points in F_1). Despite not being fully satisfactory either, these results indicate that global features which are not implemented in the CRF are of importance. This is confirmed by the CRF∩SVM setting, where CRF and SVM are stacked: This filtering procedure achieves the best precision across all models and baselines, whereas the recall is still limited. Despite being designed for exactly this purpose, the HRCRF∩SVM combination can only marginally alleviate this problem, and only at the expense of a drop in precision.

The best trade-off between precision and recall is offered by the CRF→SVM combination. This setting is not only superior to all other variants of combining a CRF with an SVM, but outperforms GNAT by 6 points in F_1 score, while being inferior to the long form baseline. However, performing feature selection on this best model using a wrapper approach (CRF→SVM+FS) leads to the overall best result of $F_1 = 0.84$, outperforming all other approaches and all baselines.

5.2.2 Individual results

Table 5 summarizes the performance of all filtering strategies broken down into individual entities. Best results are achieved for AHR, MOX and CLU. COPD forms a special case as no examples for the occurrence as a gene/protein are in the data; however the results show that the system can handle such a special distribution.

SVM and CRF are mostly outperformed by a combination of both strategies (except for CLI and HF), which shows that local and global features are highly complementary in general. Complementary cases generally favor the CRF→SVM strategy, except for PLS, where stacking is more effective.

In SAH, the pure CRF model is superior to all combinations of CRF and SVM. Apparently, the global information as contributed by the SVM is less effective than local contextual features as available to the CRF in these cases. In SAH and CLI, moreover, the best performance is obtained by the AcroTagger.

5.2.3 Impact of instance generation

All results reported in Tables 4 and 5 refer to configurations in which additional training instances have been created by manual instance generation. The impact of this method is analyzed in Table 6. The first column reports the performance of our models on the randomly sampled training data. In order to obtain the results in the second column, manual instance generation has been applied.

The results show that all our recognition models generally benefit from additional information that helps to overcome the skewed class distribution of the training data. Despite their relatively small quantity and uneven distribution across the gene/protein classes, including additional external instances yields a strong boost in all models. The largest difference is observed in SVM ($\Delta F_1 = +0.2$) and CRF→SVM ($\Delta F_1 = +0.16$). Importantly, these improvements include both precision and recall.

5.3 Feature Selection

The best feature set (*cf.* CRF→SVM+FS in Table 4) is determined by a greedy search using a wrapper approach on the best model configuration CRF→SVM. The results are depicted in Table 7. In each iteration, the table shows the best feature set detected in the previous iteration and the results for each individual feature when being added to this set. In each step, the best individual feature is kept for the next iteration. The feature analysis starts from the *long form* feature as strong baseline. The added features are, in that order, *context*, $tagger_{global}$, and $propagation_{all}$.

Overall, feature selection yields a considerable

might be seen as slightly inappropriate as the focus of GNAT is different.

[10] `ftp://ftp.ebi.ac.uk/pub/software/textmining/abbreviation_resolution/`, accessed April 23, 2014

	AHR			CLI			CLU			COPD		
Setting	P	R	F_1	P	R	F_1	P	R	F_1	P	R	F_1
SVM	1.00	0.72	0.84	0.30	0.27	0.29	1.00	0.41	0.58	0.00	1.00	0.00
CRF∩SVM	1.00	0.70	0.82	0.00	0.00	0.00	1.00	0.15	0.26	1.00	1.00	1.00
HRCRF∩SVM	1.00	0.70	0.82	1.00	0.00	0.00	1.00	0.16	0.28	1.00	1.00	1.00
CRF→SVM	0.96	0.83	0.89	0.30	0.27	0.29	1.00	0.40	0.57	0.00	1.00	0.00
CRF→SVM+FS	0.93	0.98	0.95	0.50	0.09	0.15	0.99	0.84	0.91	1.00	1.00	1.00
GNAT	0.74	0.66	0.70	1.00	0.18	0.31	0.97	0.52	0.68	1.00	1.00	1.00
CRF	0.52	0.98	0.68	0.00	0.00	0.00	1.00	0.20	0.33	0.00	1.00	0.00
AcroTagger	1.00	0.60	0.75	1.00	0.82	0.90	1.00	0.00	0.00	1.00	1.00	1.00
Long form	1.00	0.96	0.98	1.00	0.09	0.17	0.99	0.80	0.88	1.00	1.00	1.00
lex	0.40	1.00	0.57	0.05	1.00	0.09	0.89	1.00	0.94	0.00	1.00	0.00

	HF			MOX			PLS			SAH		
Setting	P	R	F_1	P	R	F_1	P	R	F_1	P	R	F_1
SVM	0.25	1.00	0.40	0.87	0.44	0.58	0.14	1.00	0.25	0.00	0.00	0.00
CRF∩SVM	1.00	0.00	0.00	1.00	0.39	0.56	1.00	1.00	1.00	1.00	0.00	0.00
HRCRF∩SVM	1.00	0.00	0.00	1.00	0.39	0.56	0.20	1.00	0.33	1.00	0.00	0.00
CRF→SVM	0.25	1.00	0.40	0.91	0.63	0.74	0.50	1.00	0.67	1.00	0.00	0.00
CRF→SVM+FS	1.00	0.00	0.00	1.00	0.37	0.54	0.00	0.00	0.00	1.00	0.00	0.00
GNAT	1.00	0.00	0.00	0.38	0.08	0.14	0.00	0.00	0.00	0.00	0.00	0.0
CRF	0.00	0.00	0.00	0.43	0.90	0.59	0.14	1.00	0.25	1.00	0.50	0.67
AcroTagger	0.33	1.00	0.50	1.00	0.00	0.00	1.00	0.00	0.00	1.00	0.60	0.75
Long form	1.00	0.00	0.00	1.00	0.00	0.00	0.00	0.00	0.00	1.00	0.00	0.00
lex	0.01	1.00	0.02	0.22	1.00	0.36	0.00	1.00	0.01	0.01	1.00	0.03

Table 5: Results for the eight genes/proteins and results for our different recognition schemes.

	randomly sampled			+instance generation		
	P	R	F_1	ΔP	ΔR	ΔF_1
SVM	0.73	0.25	0.38	+0.08	+0.20	+0.20
CRF∩SVM	1.00	0.17	0.29	-0.01	+0.09	+0.13
HRCRF∩SVM	0.97	0.18	0.30	-0.02	+0.09	+0.12
CRF→SVM	0.79	0.32	0.46	+0.05	+0.17	+0.16
CRF→SVM+FS	0.99	0.60	0.75	-0.02	+0.14	+0.09

Table 6: Impact of increasing the randomly sampled training set by adding manually curated additional positive instances (+instance generation), measured in terms of the increase in precision, recall and F_1 (ΔP, ΔR, ΔF_1).

boost in recall, while precision remains almost constant. Surprisingly, the *unigrams* feature has a particularly strong negative impact on overall performance.

While the global information contributed by the CRF turns out very valuable, accounting for most of the improvement in recall, local tagger information is widely superseded by other features. Likewise, the *abbreviation* feature does not provide any added value to the model beyond what is known from the *long form* feature.

Comparing the different feature propagation strategies, we observe that $propagation_{all}$ outperforms $propagation_{first}$.

5.4 Discussion

Our experiments show that the phenomena investigated pose a challenge to all gene recognition paradigms currently available in the literature, *i. e.*, dictionary-based, machine-learning-based (*e. g.* using a CRF), and classification-based filtering.

Our results indicate that stacking different methods suffers from a low recall in early steps of the workflow. Instead, a greedy approach that considers all occurrences of an abbreviation as input to a filtering approach yields the best performance. Incorporating information from a CRF as features into a SVM outperforms all baselines at very high levels of precision; however, the recall still leaves room for improvement.

Iter.	Feature Set	P	R	F_1	ΔF_1
1	***long form***	**0.98**	**0.65**	**0.78**	
	+*propagation*$_{1^{st}}$	0.98	0.65	0.78	+0.00
	+*propagation*$_{all}$	0.98	0.65	0.78	+0.00
	+*tagger*$_{local}$	0.72	0.81	0.76	-0.02
	+*tagger*$_{global}$	0.55	0.79	0.65	-0.13
	+context	0.98	0.67	0.79	**+0.01**
	+*abbreviation*	0.98	0.65	0.78	+0.00
	+*unigrams*	0.71	0.43	0.53	-0.25
2	***long form*** ***+context***	**0.98**	**0.67**	**0.79**	
	+*propagation*$_{1^{st}}$	0.98	0.67	0.79	+0.00
	+*propagation*$_{all}$	0.96	0.70	0.81	+0.02
	+*tagger*$_{local}$	0.98	0.70	0.82	+0.03
	+tagger$_{\mathbf{global}}$	0.97	0.72	0.83	**+0.04**
	+*abbreviation*	0.98	0.67	0.80	+0.01
	+*unigrams*	0.77	0.39	0.52	-0.27
3	***long form*** ***+context*** ***+tagger***$_{\mathbf{global}}$	**0.97**	**0.72**	**0.83**	
	+*propagation*$_{1^{st}}$	0.97	0.71	0.82	-0.01
	+propagation$_{\mathbf{all}}$	0.97	0.74	0.84	**+0.01**
	+*tagger*$_{local}$	0.97	0.72	0.82	-0.01
	+*abbreviation*	0.97	0.72	0.82	-0.01
	+*unigrams*	0.77	0.44	0.56	-0.27
4	***long form*** ***+context*** ***+tagger***$_{\mathbf{global}}$ ***+propagation***$_{\mathbf{all}}$	**0.97**	**0.74**	**0.84**	
	+*tagger*$_{local}$	0.90	0.66	0.76	-0.08
	+*abbreviation*	0.97	0.74	0.84	-0.00
	+*unigrams*	0.80	0.49	0.61	-0.23

Table 7: Greedy search for best feature combination in CRF→SVM (incl. additional positives).

In a feature selection study, we were able to show a largely positive overall impact of features that extend local contextual information as commonly applied by state-of-the-art CRF approaches. This ranges from larger context windows for collecting contextual information over abstract-level features to feature propagation strategies. However, feature selection is not equally effective in all individual classes (*cf.* Table 5).

The benefits due to feature propagation indicate that several instances of the same abbreviation in one abstract should not be considered independently of one another, although we could not verify the intuition that the first mention of an abbreviation introduces particularly valuable information for classification.

Overall, our results seem encouraging as the machinery and the features used are in general successful in determining whether an abbreviation actually denotes a gene/protein or not. The best precision/recall balance is obtained by adding CRF information as features into the classifier.

As we have shown in the cross-entity experiment setting, the system is capable of generalizing to other unseen entities. For a productive system, we assume our workflow to be applied to specific abbreviations such that the performance on other entities (and therefore on other corpora) is not substantially influenced.

6 Conclusions and Outlook

The work reported in this paper was motivated from the practical need for an effective filtering method for recognizing genes/proteins from highly ambiguous abbreviations. To the best of our knowledge, this is the first approach to tackle gene/protein recognition from ambiguous abbreviations in a systematic manner without being specific for the particular instances of ambiguous gene/protein homonyms considered.

The proposed method has been proven to allow for an improvement in recognition performance when added to an existing NER workflow. Despite being restricted to eight entities so far, our approach has been evaluated in a strict cross-entity manner, which suggests sufficient generalization power to be extended to other genes as well.

In future work, we plan to extend the data set to prove the generalizability on a larger scale and on an independent test set. Furthermore, an inclusion of the features presented in this paper into the CRF will be evaluated. Moreover, assessing the impact of the global features that turned out beneficial in this paper on other gene/protein inventories seems an interesting path to explore. Finally, we will investigate the prospects of our approach in an actual black-box evaluation setting for information retrieval.

Acknowledgements

Roman Klinger has been funded by the "It's OWL" project ("Intelligent Technical Systems Ostwestfalen-Lippe", `http://www.its-owl.de/`), a leading-edge cluster of the German Ministry of Education and Research. We thank Jörg Hakenberg and Philippe Thomas for their support in performing the baseline results with GNAT. Additionally, we thank the reviewers of this paper for their very helpful comments.

References

Hisham Al-mubaid. 2006. Biomedical term disambiguation: An application to gene-protein name disambiguation. In *In IEEE Proceedings of ITNG06*.

Judith Bandy, David Milward, and Sarah McQuay. 2009. Mining protein-protein interactions from published literature using linguamatics i2e. *Methods Mol Biol*, 563:3–13.

Steffen Bickel, Ulf Brefeld, Lukas Faulstich, Jörg Hakenberg, Ulf Leser, Conrad Plake, and Tobias Scheffer. 2004. A support vector machine classifier for gene name recognition. In *In Proceedings of the EMBO Workshop: A Critical Assessment of Text Mining Methods in Molecular Biology*.

Juliane Fluck, Heinz Theodor Mevissen, Marius Oster, and Martin Hofmann-Apitius. 2007. ProMiner: Recognition of Human Gene and Protein Names using regularly updated Dictionaries. In *Proceedings of the Second BioCreative Challenge Evaluation Workshop*, pages 149–151, Madrid, Spain.

William A. Gale, Kenneth W. Church, and David Yarowsky. 1992. One sense per discourse. In *Proceedings of the Workshop on Speech and Natural Language*, pages 233–237, Stroudsburg, PA, USA. Association for Computational Linguistics.

Sylvain Gaudan, Harald Kirsch, and Dietrich Rebholz-Schuhmann. 2005. Resolving abbreviations to their senses in medline. *Bioinformatics*, 21(18):3658–3664.

Jörg Hakenberg, Conrad Plake, Robert Leaman, Michael Schroeder, and Graciela Gonzalez. 2008. Inter-species normalization of gene mentions with GNAT. *Bioinformatics*, 24(16):i126–i132, Aug.

Jörg Hakenberg, Martin Gerner, Maximilian Haeussler, Ills Solt, Conrad Plake, Michael Schroeder, Graciela Gonzalez, Goran Nenadic, and Casey M. Bergman. 2011. The GNAT library for local and remote gene mention normalization. *Bioinformatics*, 27(19):2769–2771, Oct.

Daniel Hanisch, Juliane Fluck, Heinz-Theodor Mevissen, and Ralf Zimmer. 2003. Playing biology's name game: identifying protein names in scientific text. *Pac Symp Biocomput*, pages 403–414.

Matthias Hartung and Matthias Zwick. 2014. A corpus for the development of gene/protein recognition from rare and ambiguous abbreviations. Bielefeld University. doi:10.4119/unibi/2673424.

Antonio J Jimeno-Yepes, Bridget T McInnes, and Alan R Aronson. 2011. Exploiting mesh indexing in medline to generate a data set for word sense disambiguation. *BMC bioinformatics*, 12(1):223.

J-D. Kim, T. Ohta, Y. Tateisi, and J. Tsujii. 2003. Genia corpus–semantically annotated corpus for biotextmining. *Bioinformatics*, 19 Suppl 1:i180–i182.

Roman Klinger, Christoph M. Friedrich, Juliane Fluck, and Martin Hofmann-Apitius. 2007. Named Entity Recognition with Combinations of Conditional Random Fields. In *Proceedings of the Second BioCreative Challenge Evaluation Workshop*, Madrid, Spain, April.

Cheng-Ju Kuo, Yu-Ming Chang, Han-Shen Huang, Kuan-Ting Lin, Bo-Hou Yang, Yu-Shi Lin, Chun-Nan Hsu, and I-Fang Chung. 2007. Rich feature set, unication of bidirectional parsing and dictionary filtering for high f-score gene mention tagging. In *Proceedings of the Second BioCreative Challenge Evaluation Workshop*, Madrid, Spain, April.

Robert Leaman and Graciela Gonzalez. 2008. Banner: An executable survey of advances in biomedical named entity recognition. In Russ B. Altman, A. Keith Dunker, Lawrence Hunter, Tiffany Murray, and Teri E. Klein, editors, *Pacific Symposium on Biocomputing*, pages 652–663. World Scientific.

Vladimir I. Levenshtein. 1966. Binary codes capable of correcting deletions, insertions, and reversals. *Soviet Physics Doklady*, 10:707–710.

Tomohiro Mitsumori, Sevrani Fation, Masaki Murata, Kouichi Doi, and Hirohumi Doi. 2005. Gene/protein name recognition based on support vector machine using dictionary as features. *BMC Bioinformatics*, 6 Suppl 1:S8.

Alexander A. Morgan, Zhiyong Lu, Xinglong Wang, Aaron M. Cohen, Juliane Fluck, Patrick Ruch, Anna Divoli, Katrin Fundel, Robert Leaman, Jrg Hakenberg, Chengjie Sun, Heng-hui Liu, Rafael Torres, Michael Krauthammer, William W. Lau, Hongfang Liu, Chun-Nan Hsu, Martijn Schuemie, K Bretonnel Cohen, and Lynette Hirschman. 2008. Overview of biocreative ii gene normalization. *Genome Biol*, 9 Suppl 2:S3.

Naoaki Okazaki, Sophia Ananiadou, and Jun'ichi Tsujii. 2010. Building a high-quality sense inventory for improved abbreviation disambiguation. *Bioinformatics*, 26(9):1246–1253, May.

Martijn J. Schuemie, Ning Kang, Maarten L. Hekkelman, and Jan A. Kors. 2010. Genee: gene and protein query expansion with disambiguation. *Bioinformatics*, 26(1):147–148, Jan.

Ariel S. Schwartz and Marti A. Hearst. 2003. A simple algorithm for identifying abbreviation definitions in biomedical text. *Pac Symp Biocomput*, pages 451–462.

Burr Settles. 2005. Abner: an open source tool for automatically tagging genes, proteins and other entity names in text. *Bioinformatics*, 21(14):3191–3192, Jul.

Larry Smith, Lorraine K. Tanabe, Rie Johnson nee J. Ando, Cheng-Ju J. Kuo, I-Fang F. Chung, Chun-Nan N. Hsu, Yu-Shi S. Lin, Roman Klinger,

Christoph M. Friedrich, Kuzman Ganchev, Manabu Torii, Hongfang Liu, Barry Haddow, Craig A. Struble, Richard J. Povinelli, Andreas Vlachos, William A. Baumgartner, Lawrence Hunter, Bob Carpenter, Richard Tzong-Han T. Tsai, Hong-Jie J. Dai, Feng Liu, Yifei Chen, Chengjie Sun, Sophia Katrenko, Pieter Adriaans, Christian Blaschke, Rafael Torres, Mariana Neves, Preslav Nakov, Anna Divoli, Manuel Maña López, Jacinto Mata, and W. John Wilbur. 2008. Overview of BioCreative II gene mention recognition. *Genome biology*, 9 Suppl 2(Suppl 2):S2+.

Yoshimasa Tsuruoka and Jun'ichi Tsujii. 2003. Boosting precision and recall of dictionary-based protein name recognition. In *Proceedings of the ACL 2003 Workshop on Natural Language Processing in Biomedicine*, pages 41–48, Sapporo, Japan, July. Association for Computational Linguistics.

Hua Xu, Jung-Wei Fan, George Hripcsak, Eneida A Mendonça, Marianthi Markatou, and Carol Friedman. 2007. Gene symbol disambiguation using knowledge-based profiles. *Bioinformatics*, 23(8):1015–1022.

GuoDong Zhou, Dan Shen, Jie Zhang, Jian Su, and SoonHeng Tan. 2005. Recognition of protein/gene names from text using an ensemble of classifiers. *BMC Bioinformatics*, 6 Suppl 1:S7.

FFTM: A Fuzzy Feature Transformation Method for Medical Documents

Amir Karami, Aryya Gangopadhyay
Information Systems Department
University of Maryland Baltimore County
Baltimore, MD, 21250
amir3@umbc.edu, gangopad@umbc.edu

Abstract

The vast array of medical text data represents a valuable resource that can be analyzed to advance the state of the art in medicine. Currently, text mining methods are being used to analyze medical research and clinical text data. Some of the main challenges in text analysis are high dimensionality and noisy data. There is a need to develop novel feature transformation methods that help reduce the dimensionality of data and improve the performance of machine learning algorithms. In this paper we present a feature transformation method named FFTM. We illustrate the efficacy of our method using local term weighting, global term weighting, and Fuzzy clustering methods and show that the quality of text analysis in medical text documents can be improved. We compare FFTM with Latent Dirichlet Allocation (LDA) by using two different datasets and statistical tests show that FFTM outperforms LDA.

1 Introduction

The exponential growth of medical text data makes it difficult to extract useful information in a structured format. Some important features of text data are sparsity and high dimensionality. This means that while there may be a large number of terms in most of the documents in a corpus, any one document may contain a small percentage of those terms (Aggarwal and Zhai, 2012). This characteristic of medical text data makes feature transformation an important step in text analysis. Feature transformation is a pre-processing step in many machine-learning methods that is used to characterize text data in terms of a different number of attributes in lower dimensions. This technique has a direct impact on the quality of text mining methods. Topic models such as LDA has been used as one of popular feature transformation techniques (Ramage et al., 2010). However, fuzzy clustering methods, particularly in combination with term weighting methods, have not been explored much in medical text mining.

In this research, we propose a new method called FFTM to extract features from free-text data. The rest of the paper is organized in the following sections. In the section 2, we review related work. Section 3 contains details about our method. Section 4 describes our experiments, performance evaluation, and discussions of our results. Finally we present a summary, limitations, and future work in the last section.

2 Related Work

Text analysis is an important topic in medical informatics that is challenging due to high sparse dimensionality data. Big dimension and diversity of text datasets have been motivated medical researchers to use more feature transformation methods. Feature transformation methods encapsulate a text corpus in smaller dimensions by merging the initial features. Topic model is one of popular feature transformation methods. Among topic models, LDA (Blei et al., 2003) has been considered more due to its better performance (Ghassemi et al., 2012; Lee et al., 2010).

One of methods that has not been fully considered in medical text mining is Fuzzy clustering. Although most of Fuzzy Clusterings work in medical literature is based on image analysis (Saha and Maulik, 2014; Cui et al., 2013; Beevi and Sathik, 2012), a few work have been done in medical text mining (Ben-Arieh and Gullipalli, 2012; Fenza et al., 2012) by using fuzzy clustering. The main difference between our method and other document fuzzy clustering such as (Singh et al., 2011) is that our method use fuzzy clustering and word weighting as a pre-processing step for

Proceedings of the 2014 Workshop on Biomedical Natural Language Processing (BioNLP 2014), pages 128–133,
Baltimore, Maryland USA, June 26-27 2014.

feature transformation before implementing any classification and clustering algorithms; however, other methods use fuzzy clustering as a final step to cluster the documents. Our main contribution is to improve the quality of input data to improve the output of fuzzy clustering. Among fuzzy clustering methods, Fuzzy C-means (Bezdek, 1981) is the most popular one (Bataineh et al., 2011). In this research, we propose a novel method that combines local term weighting and global term weighting with fuzzy clustering.

3 Method

In this section, we detail our Fuzzy Feature Transformation Method *(FFTM)* and describe the steps. We begin with a brief review of LDA.

LDA is a topic model that can extract hidden topics from a collection of documents. It assumes that each document is a mixture of topics. The output of LDA are the topic distributions over documents and the word distributions over topics. In this research, we use the topics distributions over documents. LDA uses term frequency for local term weighting.

Now we introduce FFTM concepts and notations. This model has three main steps including *Local Term Weighting (LTW), Global Term Weighting (GTM), and Fuzzy Clustering* (Algorithm 1). In this algorithm, each step is the out put of each step will be the input of the next step.

Step 1: The first step is to calculate LTW. Among different LTW methods we use term frequency as a popular method. Symbol f_{ij} defines the number of times term i happens in document j.We have n documents and m words.Let

$$b(f_{ij}) = \begin{cases} 1 & f_{ij} > 0 \\ 0 & f_{ij} = 0 \end{cases} \tag{1}$$

$$p_{ij} = \frac{f_{ij}}{\sum_j f_{ij}} \tag{2}$$

The outputs of this step are $b(f_{ij})$, f_{ij} , and p_{ij}. We use them as inputs for the second step.

Step 2: The next step is to calculate GTW. We explore four GTW methods in this paper including *Entropy, Inverse Document Frequency (IDF), Probabilistic Inverse Document Frequency (ProbIDF)*, and *Normal*(Table 1).

IDF assigns higher weights to rare terms and lower weights to common terms (Papineni, 2001). ProbIDF is similar to IDF and assigns very low negative weight for the terms happen in every document (Kolda, 1998). In Entropy, it gives higher weight for the terms happen less in few documents (Dumais, 1992). Finally, Normal is used to correct discrepancies in document lengths and also normalize the document vectors. The outputs of this step are the inputs of the last step.

Step 3: Fuzzy clustering is a soft clustering technique that finds the degree of membership for each data point in each cluster, as opposed to assigning a data point only one cluster. Fuzzy clustering is a synthesis between clustering and fuzzy set theory. Among fuzzy clustering methods, Fuzzy C-means (FCM) is the most popular one and its goal is to minimize an objective func-

Algorithm 1 FFTM algorithm

Functions:E():Entropy;I():IDF;PI():ProbIDF; NO():Normal; FC():Fuzzy Clustering.
Input: Document Term Matrix
Output: Clustering membership value (μ_{ij}) for all documents and clusters.
Remove stop words
Step 1: Calculate LTW
for$i = 1$ to n**do**
  for$j = 1$ to m**do**
    Calculate f_{ij}, $b(f_{ij})$, p_{ij}
  endfor
endfor
Step 2: Calculate GTW
for$i = 1$ to n**do**
  for$j = 1$ to m**do**
    Execute E(p_{ij},n),I(f_{ij},n),PI($b(f_{ij})$,n), NO(f_{ij},n)
  endfor
endfor
Step 3: Perform Fuzzy Clustering
Execute FC(E),FC(I),FC(PI),FC(NO)

Table1: GTW Methods

Name	**Formula**
Entropy	$1 + \frac{\sum_j p_{ij} \log_2(p_{ij})}{\log_2 n}$
IDF	$\log_2 \frac{n}{\sum_j f_{ij}}$
ProbIDF	$\log_2 \frac{n - \sum_j b(f_{ij})}{\sum_j b(f_{ij})}$
Normal	$\frac{1}{\sqrt{\sum_j f_{ij}^2}}$

tion by considering constraints:

$$Min\ J_q(\mu, V, X) = \sum_{i=1}^{c}\sum_{j=1}^{n}(\mu_{ij})^q D_{ij}^2 \quad (3)$$

subject to:

$$0 \le \mu_{ij} \le 1; \quad (4)$$

$$i \in \{1, .., c\}\ and\ j \in \{1, ..., n\} \quad (5)$$

$$\sum_{i=1}^{c} \mu_{ij} = 1 \quad (6)$$

$$0 < \sum_{j=1}^{n} \mu_{ij} < n; \quad (7)$$

Where:

n= number of data
c= number of clusters
μ_{ij}= membership value
q= fuzzifier, $1 < q \le \infty$
V= cluster center vector
$D_{ij} = d(x_j, v_i)$= distance between x_j and v_i

By optimizing eq.3:

$$\mu_{ij} = \frac{1}{\sum_{k=1}^{c}(\frac{D_{ij}}{D_{kj}})^{\frac{2}{q-1}}} \quad (8)$$

$$v_i = \frac{\sum_{j=1}^{n}(\mu_{ij})^q x_j}{\sum_{j=1}^{n}(\mu_{ij})^q} \quad (9)$$

The iterations in the clustering algorithms continue till the the maximum changes in μ_{ij} becomes less than or equal to a pre-specified threshold. The computational time complexity is $O(n)$. We use μ_{ij} as the degree of clusters' membership for each document.

4 Experimental Results

In this section, we evaluate FFTM against LDA using two measures: document clustering internal metrics and document classification evaluation metrics by using one available text datasets. We use Weka[1]for classification evaluation, MALLET[2]package with its default setting for implementing LDA, Matlab fcm package[3]for implementing FCM clustering, and CVAP Matlab package[4]for clustering validation.

[1]http://www.cs.waikato.ac.nz/ml/weka/
[2]http://mallet.cs.umass.edu/
[3]http://tinyurl.com/kl33w67
[4]http://tinyurl.com/kb5bwnm

4.1 Datasets

We leverage two available datasets in this research. Our first test dataset called Deidentified Medical Text[5] is an unlabeled corpus of 2434 nursing notes with 12,877 terms after removing stop words. The second dataset [6] is a labeled corpus of English scientific medical abstracts from Springer website. It is included 41 medical journals ranging from Neurology to Radiology. In this research, we use the first 10 journals including: Arthroscopy, Federal health standard sheet, The anesthetist, The surgeon, The gynecologist, The dermatologist, The internist, The neurologist, The Ophthalmology, The orthopedist, and The pathologist. In our experiments we select three subsets from the above journals, the first two with 4012 terms and 171 documents, first five with 14189 terms and 1527 documents, and then all ten respectively with 23870 terms and 3764 documents to track the performance of FFTM and LDA by increasing the number of documents and labels.

4.2 Document Clustering

The first evaluation comparing FFTM with LDA is document clustering by using the first dataset. Internal and external validation are two major methods for clustering validation; however, comparison between these two major methods shows that internal validation is more more precise (Rendón et al., 2011). We evaluate different number of features (topics) and clusters by using two internal clustering validation methods including Silhouette index and Calinski-Harabasz index using K-means with 500 iterations. Silhouette index shows that how closely related are objects in a cluster and how distinct a cluster from other other clusters. The higher value means the better result.The Silhouette index (S) is defined as:

$$S(i) = \frac{(b(i) - a(i))}{Max\{a(i), b(i)\}} \quad (10)$$

Where $a(i)$ is the average dissimilarity of sample i with the same data in a cluster and $b(i)$ is the minimum average dissimilarity of sample i with other data that are not in the same cluster.

Calinski-Harabasz index (CH) valuates the cluster validity based on the average between- and within-cluster sum of squares.It is defined as:

[5]http://tinyurl.com/kfz2hm4
[6]http://tinyurl.com/m2c8se6

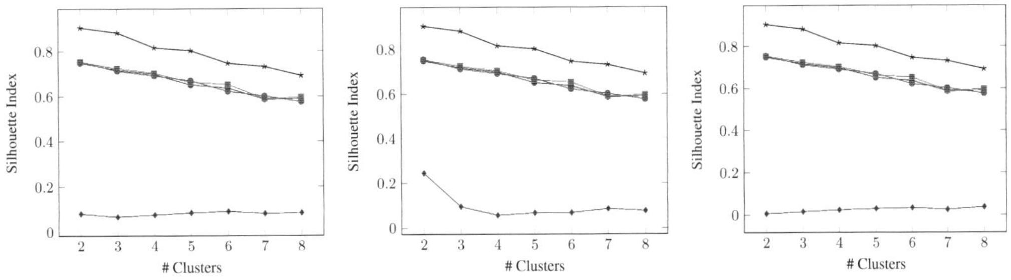

(a) 50 Features (b) 100 Features (c) 150 Features

Figure1: Clustering Validation with Silhouette Index

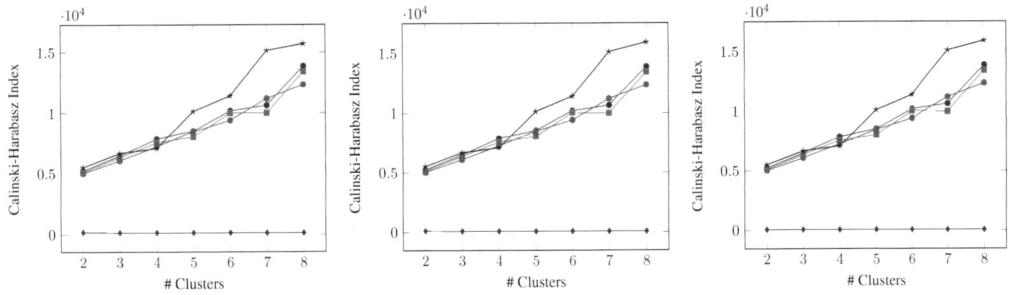

(a) 50 Features (b) 100 Features (c) 150 Features

Figure2: Clustering Validation with Calinski-Harabasz Index

$$CH = \frac{trace(S_B)}{trace(S_W)} \cdot \frac{n_p - 1}{n_p - k} \quad (11)$$

Where (S_B) is the between-cluster scatter matrix, (S_W) the internal scatter matrix, n_p is the number of clustered samples, and k is the number of clusters. Higher value indicates a better clustering. We track the performance of both FFTM and LDA using different number of clusters ranging from 2 to 8 with different number of features including 50, 100, and 150. Both Silhouette index and Calinski-Harabasz index show that FFTM is the best method with all ranges of features and clusters (Figures 1 and 2). The gap between FFTM and LDA does not change a lot by using different number of features and clusters. LDA has the lowest performance and Normal has the best performance among GTW methods in different ranges of features and clusters. According to the paired difference test, the improvement of FFTM over LDA is statistically significant with a $p - value < 0.05$ using the two internal clustering validation methods.

4.3 Document Classification

The second evaluation measure is document classification by using the second datasest. We evaluate different number of classes and features (topics) with accuracy, F-measure, and ROC using Random Forest. Accuracy is the portion of true results in a dataset. F-measure is another measure of classification evaluation that considers both precision and recall. ROC curves plot False Positive on the X axis vs. True Positive on the Y axis to find the trade off between them; therefore, the closer to the upper left indicates better performance. We assume more documents and classes have more topics;therefore, we choose 100 features for two classes, 150 features for five classes, and 200 features for ten classes. In addition, we use 10 cross validation as test option.

This experiment shows that FFTM has the best performance in different number of features and labels (Table 2). LDA has the lowest performance and the average performance of ProbIDF has the best among GTW methods in all ranges of features and clusters. According to the paired difference test, the improvement of FFTM over LDA is statistically significant with a $p - value < 0.05$.

Table2: The Second Dataset Classification Performance

Method	#Features	# Labels	Acc %	F-Measure	ROC
FFTM(Entropy)	100	2	96.49	0.959	0.982
FFTM(IDF)	100	2	98.24	0.982	0.996
FFTM(ProIDF)	100	2	97.66	0.977	0.987
FFTM(Normal)	100	2	92.39	0.912	0.971
LDA	100	2	90.06	0.9	0.969
FFTM(Entropy)	150	5	71.84	0.694	0.874
FFTM(IDF)	150	5	70.79	0.686	0.859
FFTM(ProIDF)	150	5	70.39	0.674	0.859
FFTM(Normal)	150	5	68.11	0.649	0.851
LDA	150	5	66.27	0.637	0.815
FFTM(Entropy)	200	10	51.06	0.501	0.828
FFTM(IDF)	200	10	51.73	0.506	0.826
FFTM(ProIDF)	200	10	53.72	0.525	0.836
FFTM(Normal)	200	10	50.05	0.485	0.815
LDA	200	10	47.68	0.459	0.792

5 Conclusion

The explosive growth of medical text data makes text analysis as a key requirement to find patterns in datasets;however, the typical high dimensionality of such features motivates researchers to utilize dimension reduction techniques such as LDA. Although LDA has been considered more recently in medical text analysis (Jimeno-Yepes et al., 2011), fuzzy clustering methods such as FCM has not been used in medical text clustering, but rather in image processing. In the current study, we propose a method called FFTM to combine LTW and GTM with Fuzzy clustering, and compare its performance with that of LDA. We use different sets of data including different number of features, different number of clusters, and different number of classes.The findings of this study show that combining FCM with LTW and GTW methods can significantly improve medical documents analysis. We conclude that different factors including number of features, number of clusters, and classes can affect the outputs of machine learning algorithms. In addition, the performance of FFTM is improved by using GTW methods. This method proposed in this paper may be applied to other medical documents to improve text analysis outputs. One limitation of this paper is that we use one clustering method, one classification method, and two internal clustering validation methods for evaluation. Our future direction is to explore more machine learning algorithms and clustering validation methods for evaluation and also other fuzzy clustering algorithms for feature transformation. The main goal of future research is to present an efficient and effective medical topic model using fuzzy set theory.

References

CharuC Aggarwal and ChengXiang Zhai. 2012. An introduction to text mining. In *Mining Text Data*, pages 1–10. Springer.

KMBataineh, MNaji, and MSaqer. 2011. A comparison study between various fuzzy clustering algorithms. *Jordan Journal of Mechanical & Industrial Engineering*, 5(4).

Zulaikha Beevi and Mohamed Sathik. 2012. A robust segmentation approach for noisy medical images using fuzzy clustering with spatial probability. *International Arab Journal of Information Technology (IAJIT)*, 9(1).

David Ben-Arieh and DeepKumar Gullipalli. 2012. Data envelopment analysis of clinics with sparse data: Fuzzy clustering approach. *Computers & Industrial Engineering*, 63(1):13–21.

JamesC Bezdek. 1981. *Pattern recognition with fuzzy objective function algorithms*. Kluwer Academic Publishers.

DavidM Blei, AndrewY Ng, and MichaelI Jordan. 2003. Latent dirichlet allocation. *the Journal of machine Learning research*, 3:993–1022.

Wenchao Cui, YiWang, Yangyu Fan, Yan Feng, and Tao Lei. 2013. Global and local fuzzy clustering with spatial information for medical image segmentation. In *Signal and Information Processing (ChinaSIP), 2013 IEEE China Summit & International Conference on*, pages 533–537. IEEE.

Susan Dumais. 1992. Enhancing performance in latent semantic indexing (lsi) retrieval.

Giuseppe Fenza, Domenico Furno, and Vincenzo Loia. 2012. Hybrid approach for context-aware service discovery in healthcare domain. *Journal of Computer and System Sciences*, 78(4):1232–1247.

Marzyeh Ghassemi, Tristan Naumann, Rohit Joshi, and Anna Rumshisky. 2012. Topic models for mortality modeling in intensive care units. In *ICML Machine Learning for Clinical Data Analysis Workshop*.

Antonio Jimeno-Yepes, Bartłomiej Wilkowski, JamesG Mork, Elizabeth VanLenten, DinaDemner Fushman, and AlanR Aronson. 2011. A bottom-up approach to medline indexing recommendations. In *AMIA Annual Symposium Proceedings*, volume 2011, page 1583. American Medical Informatics Association.

TamaraG Kolda. 1998. Limited-memory matrix methods with applications.

Sangno Lee, Jeff Baker, Jaeki Song, and JamesC Wetherbe. 2010. An empirical comparison of four text mining methods. In *System Sciences (HICSS), 2010 43rd Hawaii International Conference on*, pages 1–10. IEEE.

Kishore Papineni. 2001. Why inverse document frequency? In *Proceedings of the second meeting of the North American Chapter of the Association for Computational Linguistics on Language technologies*, pages 1–8. Association for Computational Linguistics.

Daniel Ramage, SusanT Dumais, and DanielJ Liebling. 2010. Characterizing microblogs with topic models. In *ICWSM*.

Eréndira Rendón, Itzel Abundez, Alejandra Arizmendi, and ElviaM Quiroz. 2011. Internal versus external cluster validation indexes. *International Journal of computers and communications*, 5(1):27–34.

Indrajit Saha and Ujjwal Maulik. 2014. Multiobjective differential evolution-based fuzzy clustering for mr brain image segmentation image segmentation. In *Advanced Computational Approaches to Biomedical Engineering*, pages 71–86. Springer.

VivekKumar Singh, Nisha Tiwari, and Shekhar Garg. 2011. Document clustering using k-means, heuristic k-means and fuzzy c-means. In *Computational Intelligence and Communication Networks (CICN), 2011 International Conference on*, pages 297–301. IEEE.

Using statistical parsing to detect agrammatic aphasia

Kathleen C. Fraser[1], Graeme Hirst[1], Jed A. Meltzer[2], Jennifer E. Mack[3], and Cynthia K. Thompson[3,4,5]

[1]Dept. of Computer Science, University of Toronto
[2]Rotman Research Institute, Baycrest Centre, Toronto
[3]Dept. of Communication Sciences and Disorders, Northwestern University
[4]Dept. of Neurology, Northwestern University
[4]Cognitive Neurology and Alzheimer's Disease Center, Northwestern University

{kfraser,gh}@cs.toronto.edu, jmeltzer@research.baycrest.org
{jennifer-mack-0,ckthom}@northwestern.edu

Abstract

Agrammatic aphasia is a serious language impairment which can occur after a stroke or traumatic brain injury. We present an automatic method for analyzing aphasic speech using surface level parse features and context-free grammar production rules. Examining these features individually, we show that we can uncover many of the same characteristics of agrammatic language that have been reported in studies using manual analysis. When taken together, these parse features can be used to train a classifier to accurately predict whether or not an individual has aphasia. Furthermore, we find that the parse features can lead to higher classification accuracies than traditional measures of syntactic complexity. Finally, we find that a minimal amount of pre-processing can lead to better results than using either the raw data or highly processed data.

1 Introduction

After a stroke or head injury, individuals may experience aphasia, an impairment in the ability to comprehend or produce language. The type of aphasia depends on the location of the lesion. However, even two patients with the same type of aphasia may experience different symptoms. A careful analysis of narrative speech can reveal specific patterns of impairment, and help a clinician determine whether an individual has aphasia, what type of aphasia it is, and how the symptoms are changing over time.

In this paper, we present an automatic method for the analysis of one type of aphasia, *agrammatic aphasia*. characterized by the omission of function words, the omission or substitution of morphological markers for person and number, the absence of verb inflection, and a relative increase in the number of nouns and decrease in the number of verbs (Bastiaanse and Thompson, 2012). There is often a reduction in the variety of different syntactic structures used, as well as a reduction in the complexity of those structures (Progovac, 2006). There may also be a strong tendency to use the canonical word order of a language, for example subject-verb-object in English (Progovac, 2006).

Most studies of narrative speech in agrammatic aphasia are based on manually annotated speech transcripts. This type of analysis can provide detailed and accurate information about the speech patterns that are observed. However, it is also very time consuming and requires trained transcribers and annotators. Studies are necessarily limited to a manageable size, and the level of agreement between annotators can vary.

We propose an automatic approach that uses information from statistical parsers to examine properties of narrative speech. We extract context-free grammar (CFG) production rules as well as phrase-level features from syntactic parses of the speech transcripts. We show that this approach can detect many features which have been previously reported in the aphasia literature, and that classification of agrammatic patients and controls can be achieved with high accuracy.

We also examine the effects of including speech dysfluencies in the transcripts. Dysfluencies and non-narrative words are usually removed from the transcripts as a pre-processing step, but we show that by retaining some of these items, we can actually achieve a higher classification accuracy than by using the completely clean transcripts.

Finally, we investigate whether there is any benefit to using the parse features instead of more traditional measures of syntactic complexity, such as Yngve depth or mean sentence length. We find that the parse features convey more information

Proceedings of the 2014 Workshop on Biomedical Natural Language Processing (BioNLP 2014), pages 134–142,
Baltimore, Maryland USA, June 26-27 2014. ©2014 Association for Computational Linguistics

about the specific syntactic structures being produced (or avoided) by the agrammatic speakers, and lead to better classification accuracies.

2 Related Work

2.1 Syntactic analysis of agrammatic narrative speech

Much of the previous work analyzing narrative speech in agrammatic aphasia has been performed manually. One widely used protocol is called Quantitative Production Analysis (QPA), developed by Saffran et al. (1989). QPA can be used to measure morphological content, such as whether determiners and verb inflections are produced in obligatory contexts, as well as structural complexity, such as the number of embedded clauses per sentence. Subsequent studies have found a number of differences between normal and agrammatic speech using QPA (Rochon et al., 2000). Another popular protocol called the Northwestern Narrative Language Analysis (NNLA) was introduced by Thompson et al. (1995). This protocol analyzes each utterance at five different levels, and focuses in particular on the production of verbs and verb argument structure.

Perhaps more analogous to our work here, Goodglass et al. (1994) conducted a detailed examination of the syntactic constituents used by aphasic patients and controls. In that study, utterances were grouped according to how many syntactic constituents they contained. They found that agrammatic participants were more likely to produce single-constituent utterances, especially noun phrases, and less likely to produce subordinate clauses. They also found that agrammatic speakers sometimes produced two-constituent utterances consisting of only a subject and object, with no verb. This pattern was never observed in control speech.

A much smaller body of work explores the use of computational techniques to analyze agrammatism. Holmes and Singh (1996) analyzed conversational speech from aphasic speakers and controls. Their features mostly included measures of vocabulary richness and frequency counts of various parts-of-speech (e.g. nouns, verbs); however they also measured "clause-like semantic unit rate". This feature was intended to measure the speaker's ability to cluster words together, although it is not clear what the criteria for segmenting clause-like units were or whether it was done manually or automatically. Nonetheless, it was found to be one of the most important variables for distinguishing between patients and controls.

MacWhinney et al. (2011) presented several examples of how researchers can use the AphasiaBank[1] database and associated software tools to conduct automatic analyses (although the transcripts are first hand-coded for errors by experienced speech-language pathologists). Specifically with regards to syntax, they calculated several frequency counts and ratios for different parts-of-speech and bound morphemes. There was one extension beyond treating each word individually: this involved searching for pre-defined collocations such as *once upon a time* or *happily ever after*, which were found to occur more rarely in the patient transcripts than in the control transcripts.

We present an alternative, automated method of analysis. We do not attempt to fully replicate the results of the manual studies, but rather provide a complementary set of features which can indicate grammatic abnormalities. Unlike previous computational studies, we attempt to move beyond single-word analysis and examine which patterns of syntax might indicate agrammatism.

2.2 Using parse features to assess grammaticality

Syntactic complexity metrics derived from parse trees have been used by various researchers in studies of mild cognitive impairment (Roark et al., 2011), autism (Prud'hommeaux et al., 2011), and child language development (Sagae et al., 2005; Hassanali et al., 2013). Here we focus specifically on the use of CFG production rules as features.

Using the CFG production rules from statistical parsers as features was first proposed by Baayen et al. (1996), who applied the features to an authorship attribution task. More recently, similar features have been widely used in native language identification (Wong and Dras, 2011; Brooke and Hirst, 2012; Swanson and Charniak, 2012). Perhaps most relevant to the task at hand, CFG productions as well as other parse outputs have proved useful for judging the grammaticality and fluency of sentences. For example, Wong and Dras (2010) used CFG productions to classify sentences from an artificial error corpus as being either grammatical or ungrammatical.

Taking a different approach, Chae and Nenkova

[1] `http://talkbank.org/AphasiaBank/`

	Agrammatic ($N = 24$)	Control ($N = 15$)
Male/Female	15/9	8/7
Age (years)	58.1 (10.6)	63.3 (6.4)
Education (years)	16.3 (2.5)	16.4 (2.4)

Table 1: Demographic information. Numbers are given in the form: mean (standard deviation).

(2009) calculated several surface features based on the output of a parser, such as the length and relative proportion of different phrase types. They used these features to distinguish between human and machine translations, and to determine which of a pair of translations was the more fluent. However, to our knowledge there has been no work using parser outputs to assess the grammaticality of speech from individuals with post-stroke aphasia.

3 Data

3.1 Participants

This was a retrospective analysis of data collected by the the Aphasia and Neurolinguistics Research Laboratory at Northwestern University. All agrammatic participants had experienced a stroke at least 1 year prior to the narrative sample collection. Demographic information for the participants is given in Table 1. There is no significant ($p < 0.05$) difference between the patient and control groups on age or level of education.

3.2 Narrative task

To obtain a narrative sample, the participants were asked to relate the well-known fairy tale *Cinderella*. Each participant was first given a wordless picture book of the story to look through. The book was then removed, and the participant was asked to tell the story in his or her own words. The examiner did not interrupt or ask questions.

The narratives were recorded and later transcribed following the NNLA protocol. The data was segmented into utterances based on syntactic and prosodic cues. Filled pauses, repetitions, false starts, and revisional phrases (e.g. *I mean*) were all placed inside parentheses. The average length of the raw transcripts was 332 words for agrammatic participants and 387 words for controls; when the non-narrative words were excluded the average length was 194 words for the agrammatic group and 330 for controls.

4 Methods

4.1 Parser Features

We consider two types of features: CFG production rules and phrase-level statistics. For the CFG production rules, we use the Charniak parser (Charniak, 2000) trained on Wall Street Journal data to parse each utterance in the transcript and then extract the set of non-lexical productions. The total number of types of productions is large, many of them occurring very infrequently, so we compile a list of the 50 most frequently occurring productions in each of the two groups (agrammatic and controls) and use the combined set as the set of features. The feature values can be binary (does a particular production rule appear in the narrative or not?) or integer (how many times does a rule occur?). The CFG non-terminal symbols follow the Penn Treebank naming conventions.

For our phrase-level statistics, we use a subset of the features described by Chae and Nenkova (2009), which are related to the incidence of different phrase types. We consider three different phrase types: noun phrases, verb phrases, and prepositional phrases. These features are defined as follows:

- *Phrase type proportion*: Length of each phrase type (including embedded phrases), divided by total narrative length.
- *Average phrase length*: Total number of words in a phrase type, divided by number of phrases of that type.
- *Phrase type rate*: Number of phrases of a given type, divided by total narrative length.

Because we are judging the grammaticality of the entire narrative, we normalize by narrative length (rather than sentence length, as in Chae and Nenkova's study). These features are real-valued.

We first perform the analysis on the transcribed data with the dysfluencies removed, labeled the "clean" dataset. This is the version of the transcript that would be used in the manual NNLA analysis. However, it is the result of human effort and expertise. To test the robustness of the system on data that has not been annotated in this way, we also use the "raw" dataset, with no dysfluencies removed (i.e. including everything inside the parentheses), and an "auto-cleaned" dataset, in which filled pauses are automatically removed from the raw transcripts. We also use a simple algorithm to remove "stutters" and false starts, by removing non-word tokens of length one or two (e.g. *C- C- Cinderella* would become simply *Cinderella*). This provides a more realistic view of the performance of our system on real data. We also hypothesize that there may be important information to be found in the dysfluent speech segments.

4.2 Feature weighting and selection

We assume that some production rules will be more relevant to the classification than others, and so we want to weight the features accordingly. Using *term frequency–inverse document frequency* (*tf-idf*) would be one possibility; however, the *tf-idf* weights do not take into account any class information. *Supervised term weighting* (STW), has been proposed by Debole and Sebastiani (2004) as an alternative to *tf-idf* for text classification tasks. In this weighting scheme, feature weights are assigned using the same algorithm that is used for feature selection. For example, one way to select features is to rank them by their information gain (InfoGain). In STW, the InfoGain value for each feature is also used to replace the *idf* term. This can be expressed as $W(i,d) = \mathrm{df}(i,d) \times \mathrm{InfoGain}(i)$, where $W(i,d)$ is the weight assigned to feature i in document d, $\mathrm{df}(i,d)$ is the frequency of occurrence of feature i in document d, and $\mathrm{InfoGain}(i)$ is the information gain of feature i across all the training documents.

We considered two different methods of STW: weighting by InfoGain and weighting by gain ratio (GainRatio). The methods were also used as feature selection, since any feature that was assigned a weight of zero was removed from the classification. We also consider *tf-idf* weights and unweighted features for comparison.

4.3 Syntactic complexity metrics

To compare the performance of the parse features with more-traditional syntactic complexity metrics (SC metrics), we calculate the mean length of utterance (MLU), mean length of T-unit[2] (MLT), mean length of clause (MLC), and parse tree height. We also calculate the mean, maximum, and total Yngve depth, which measures the proportion of left-branching to right-branching in each parse tree (Yngve, 1960). These measures are commonly used in studies of impaired language (e.g. Roark et al. (2011), Prud'hommeaux et al. (2011), Fraser et al. (2013b)). We hypothesize that the parse features will capture more information about the specific impairments seen in agrammatic aphasia; however, using the general measures of syntactic complexity may be sufficient for the classifiers to distinguish between the groups.

4.4 Classification

To test whether the features can effectively distinguish between the agrammatic group and controls, we use them to train and test a machine learning classifier. We test three different classification algorithms: naive Bayes (NB), support vector machine (SVM), and random forests (RF). We use a leave-one-out cross-validation framework, in which one transcript is held out as a test set, and the other transcripts form the training data. The feature weights are calculated on the training set and then applied to the test set (as a result, each fold of training/testing may use different features and feature weights). The SVM and RF algorithms are tuned in a nested cross-validation loop. The classifier is then tested on the held-out point. This procedure is repeated across all data points, and the average accuracy is reported.

A baseline classifier which assigns all data to the largest class would achieve an accuracy of .62 on this classification task. For a more realistic measure of performance, we also compare our results to the baseline accuracy that can be achieved using only the length of the narrative as input.

5 Results

5.1 Features using clean transcripts

We first present the results for the clean transcripts. Although different features may be selected in each fold of the cross-validation, for simplicity we show only the feature rankings on the whole data set. Table 2 shows the top features as ranked by GainRatio. The frequencies are given to indicate the direction of the trend; they represent the average frequency per narrative for each class (agrammatic = AG and control = CT). Boldface indicates the group with the higher frequency. Asterisks are used to indicate the significance of the difference between the groups.

When working with clinical data, careful examination of the features can be beneficial. By comparing features with previous findings in the literature on agrammatism, we can be confident that we are measuring real effects and not just artifacts

[2] A T-unit consists of a main clause and its attached dependent clauses.

	Rule	AG freq	CT freq	p
1	PP → IN NP	10.3	**24.9**	***
2	ROOT → NP	**2.9**	0.2	***
3	NP → DT NN POS	0.0	**0.7**	*
4	NP → PRP$ JJ NN	0.5	**0.7**	*
5	VP → TO VP	4.2	**7.5**	*
6	NP → NNP	5.9	**6.6**	
7	VP → VB PP	1.1	**2.9**	**
8	VP → VP CC VP	1.1	**3.1**	**
9	NP → DT NN NN	1.0	**2.7**	**
10	VP → VBD VP	0.1	**0.5**	*
11	WHADVP → WRB	0.5	**1.4**	*
12	FRAG → NP .	**0.7**	0.0	**
13	NP → JJ NN	**0.7**	0.0	**
14	SBAR → WHNP S	1.7	**3.1**	*
15	NP → NP SBAR	1.6	**2.5**	
16	S → NP VP	7.8	**16.1**	**
17	NP → PRP$ JJ NNS	0.0	**0.5**	*
18	NP → PRP$ NN NNS	0.0	**0.6**	*
19	SBAR → WHADVP S	0.4	**1.2**	*
20	VP → VBN PP	0.4	**2.0**	*

Table 2: Top 20 features ranked by GainRatio using the clean transcripts. ($^{*}p < 0.05$, $^{**}p < 0.005$, $^{***}p < 0.0005$).

of the parsing algorithm. This can also potentially provide an opportunity to observe features of agrammatic speech that have not been examined in manual analyses. We examine the top-ranked features in Table 2 in some detail, especially as they relate to previous work on agrammatism. In particular, the top features suggest some of the following features of agrammatic speech:

- Reduced number of prepositional phrases. This is suggested by feature 1, PP → IN NP. It is also reflected in features 7 and 20.
- Impairment in using verbs. We can see in feature 2 (ROOT → NP) that there is a greater number of utterances consisting of only a noun phrase. Feature 12 is also consistent with this pattern (FRAG → NP .). We also observe a reduced number of coordinated verb phrases (VP → VP CC VP).
- Omission of grammatical morphemes and function words. The agrammatic speakers use fewer possessives (NP → DT NN POS). Feature 9 indicates that the control participants more frequently produce compound nouns with a determiner (often ***the glass slipper*** or ***the fairy godmother***). Feature 4 also suggests some difficulty with determiners, as the agrammatic participants produce fewer nouns modified by a possessive pronoun and an adjective. Contrast this with feature 13, which shows agrammatic speech is more likely to contain noun phrases containing just an adjective and a noun. For example, in the control narratives we are more likely to see phrases such as *her godmother ... waves* ***her magic wand***, while in the agrammatic narratives phrases like *Cinderella had* ***wicked stepmother*** are more common.
- Reduced number of embedded clauses and phrases. Evidence for this can be found in the reduced number of wh-adverb phrases (WHADVP → WRB), as well as features 14, 15, and 19.

	NB	SVM	RF
Narrative length	.62	.56	.64
Binary, no weights	**.87**	**.87**	.77
Binary, *tf-idf*	.87	**.90**	.85
Binary, InfoGain	.82	**.90**	.74
Binary, GainRatio	**.90**	.82	.79
Frequency, no weights	**.90**	.85	.85
Frequency, *tf-idf*	**.85**	.82	.77
Frequency, InfoGain	**.90**	**.90**	.82
Frequncy, GainRatio	.90	**.92**	.74
SC metrics, no weights	**.85**	.77	.82
SC metrics, InfoGain	**.85**	.77	.79
SC metrics, GainRatio	**.85**	.77	.82

Table 3: Average classification accuracy using the clean transcripts. The highest classification accuracy for each feature set is indicated with boldface.

The results of our classification experiment on the clean data are shown in Table 3. The results are similar for the binary and frequency features, with the best result of .92 achieved using an SVM classifier and frequency features, with GainRatio weights. The best results using parse features (.85–.92) are the same or slightly better than the best results using SC features (.85), and both feature sets perform above baseline.

5.2 Effect of non-narrative speech

In this section we perform two additional experiments, using the raw and auto-cleaned transcripts.

	Rule	AG freq.	CT freq.	p
1	NP → DT NN POS	0.0	**0.5**	*
2	PP → IN NP	12.2	**26.1**	***
3	SBAR → WHADVP S	0.4	**1.5**	*
4	VP → VBD	0.75	**1.1**	
5	VP → TO VP	4.3	**7.3**	*
6	S → CC PP NP VP .	0.04	**0.5**	*
7	NP → PRP$ JJ NNS	0.04	**0.5**	*
8	VP → AUX VP	3.7	**6.0**	
9	ROOT → FRAG	**4.5**	0.7	**
10	ADVP → RB	9.8	**12.3**	
11	NP → NNP	4.4	**6.2**	*
12	NP → DT NN	15.0	**24.1**	**
13	VP → VB PP	1.2	**2.8**	*
14	VP → VP CC VP	1.0	**2.9**	*
15	WHADVP → WRB	0.6	**1.5**	*
16	VP → VBN PP	0.4	**2.0**	*
17	INTJ → UH UH	**3.5**	0.3	*
18	VP → VBP NP	**0.5**	0.0	*
19	NP → NNP NNP	**1.5**	0.5	**
20	S → CC ADVP NP VP .	1.3	**2.3**	

Table 4: Top 20 features ranked by GainRatio using the raw transcripts. Bold feature numbers indicate rules which did not appear in Table 2. ($^*p < 0.05$, $^{**}p < 0.005$, $^{***}p < 0.0005$).

We discuss the differences between the selected features in each case, and the resulting classification accuracies.

Using the raw transcripts, we find that the ranking of features is markedly different than with the human-annotated transcripts (Table 4, bold feature numbers). Examining these production rules more closely, we observe some characteristics of agrammatic speech which were not detectable in the annotated transcripts:

- Increased number of dysfluencies. We observe a higher number of consecutive fillers (INTJ → UH UH) in the agrammatic data, as well as a higher number of consecutive proper nouns (NP → NNP NNP), usually two attempts at Cinderella's name. Feature 18 (VP → VBP NP) also appears to support this trend, although it is not immediately obvious. Most of the control participants tell the story in the past tense, and if they do use the present tense then the verbs are often in the third-person singular (*Cinderella **finds her fairy godmother***). Looking at the data, we found that feature 18 can indicate a verb agreement error, as in *he **attend the ball***. However, in almost twice as many cases it indicates use of the discourse markers *I mean* and *you know*, followed by a repaired or target noun phrase.
- Decreased connection between sentences. Feature 6 shows a canonical NP VP sentence, preceded by a coordinate conjunction and a prepositional phrase. Some examples of this from the control transcripts include, *And at the stroke of midnight*... and *And in the process* .... The conjunction creates a connection from one utterance to the next, and the prepositional phrase indicates the temporal relationship between events in the story, creating a sense of cohesion. See also the similar pattern in feature 20, representing sentence beginnings such as *And then* ....

However, there are some features which were highly ranked in the clean transcripts but do not appear in Table 4. What information are we losing by using the raw data? One issue with using the raw transcripts is that the inclusion of filled pauses "splits" the counts for some features. For example, the feature FRAG → NP . is ranked 12th using the clean transcripts but does not appear in the top 20 when using the raw transcripts. When we examine the transcripts, we find that the phrases that are counted in this feature in the clean transcripts are actually split into three features in the raw transcripts: FRAG → NP ., FRAG → INTJ NP ., and FRAG → NP INTJ ..

The classification results for the raw transcripts are given in Table 5. The results are similar to those for the clean transcripts, although in this case the best accuracy (.92) is achieved in three different configurations (all using the SVM classifier). The phrase-level features out-perform the traditional SC measures in only half the cases.

Using the auto-cleaned transcripts, we see some similarities with the previous cases (Table 6). However, some of the highly ranked features which disappeared when using the raw transcripts are now significant again (e.g. ROOT → NP, FRAG → NP .). There are also three remaining features which are significant and have not yet been discussed. Feature 9 shows an increased use of determiners with proper nouns (e.g. *the Cinderella*), a frank grammatical error. Feature 20

	NB	SVM	RF
Narrative length	.51	.62	.69
Binary, no weights	.87	**.92**	.82
Binary, *tf-idf*	.87	**.92**	.72
Binary, InfoGain	.85	**.87**	.82
Binary, GainRatio	.82	**.87**	.85
Frequency, no weights	.85	**.90**	.69
Frequency, *tf-idf*	.82	**.92**	.90
Frequency, InfoGain	**.85**	.74	**.85**
Frequncy, GainRatio	**.85**	.74	.82
SC metrics, no weights	.74	.79	**.82**
SC metrics, InfoGain	.77	**.85**	**.85**
SC metrics, GainRatio	.77	.85	**.87**

Table 5: Average classification accuracy using raw transcripts. The highest classification accuracy for each feature set is indicated with boldface.

	Rule	AG freq.	CT freq.	p
1	PP → IN NP	12.0	**26.0**	***
2	NP → DT NN POS	0.0	**0.7**	*
3	VP → VP CC VP	0.8	**2.9**	**
4	S → CC SBAR NP VP .	0.0	**0.5**	
5	SBAR → WHADVP S	0.4	**1.5**	*
6	NP → NNP	5.6	**6.7**	
7	VP → VBD	0.8	**1.1**	
8	S → CC PP NP VP .	0.04	**0.6**	*
9	NP → DT NNP	**0.6**	0.0	**
10	VP → TO VP	4.6	**7.5**	*
11	ROOT → FRAG	**3.0**	0.5	***
12	ROOT → NP	**2.1**	0.1	*
13	VP → VBP NP	1.7	**3.6**	
14	NP → PRP$ JJ NNS	0.04	**0.5**	*
15	VP → VB PP	1.1	**2.8**	**
16	VP → VBN PP	0.4	**1.9**	*
17	FRAG → NP .	**0.4**	0.0	*
18	NP → NNP .	**2.1**	0.1	
19	S → NP ADJP	**0.4**	0.0	*
20	FRAG → CC NP .	**0.7**	0.07	**

Table 6: Top 10 features ranked by GainRatio using the auto-cleaned transcripts. Bold feature numbers indicate rules which did not appear in Table 2. ($^{*}p < 0.05$, $^{**}p < 0.005$, $^{***}p < 0.0005$).

provides another example of a sentence fragment with no verb. Finally, feature 19 represents an increased number of sentences or clauses consisting of a noun phrase followed by adjective phrase. Looking at the transcripts, this is not generally indicative of an error, but rather use of the word *okay*, as in *she dropped **her shoe okay***.

The classification results for the auto-cleaned data, shown in Table 7, show a somewhat different pattern from the previous experiments. The accuracies using the parse features are generally higher, and the best result of .97 is achieved using the binary features and the naive Bayes classifier. Interestingly, this data set also results in the lowest accuracy for the syntactic complexity metrics.

5.3 Phrase-level parse features

The classifiers in Tables 3, 5, and 7 used the phrase-level parse features as well as the CFG productions. Although these features were calculated for NPs, VPs, and PPs, the NP features were never selected by the GainRatio ranking algorithm, and did not differ significantly between groups. The significance levels of the VP and PP features are reported in Table 8. PP rate and proportion are significantly different in all three sets of transcripts, which is consistent with the high ranking of PP → IN NP in each case. VP rate and proportion are often significant, although less so. Notably, PP and VP length are both significant in the clean transcripts, but not significant in the raw transcripts and only barely significant in the auto-cleaned transcripts.

5.4 Analysis of variance

With a multi-way ANOVA we found significant main effects of classifier ($F(2,63) = 11.6$, $p < 0.001$) and data set ($F(2,63) = 11.2$, $p < 0.001$) on accuracy. A Tukey post-hoc test revealed significant differences between SVM and RF ($p < 0.001$) and NB and RF ($p < 0.001$) but not between SVM and NB. As well, we see a significant difference between the clean and auto-cleaned data ($p < 0.001$) and the raw and auto-cleaned data ($p < 0.001$) but not between the raw and clean data. There was no significant main effect of weighting scheme or feature type (binary or frequency) on accuracy. We did not examine any possible interactions between these variables.

6 Discussion

6.1 Transcripts

We achieved the highest classification accuracies using the auto-cleaned transcripts. The raw transcripts, while containing more information about dysfluent events, also seemed to cause more dif-

	NB	SVM	RF
Narrative length	.51	.62	.64
Binary, no weights	.92	**.95**	.90
Binary, *tf-idf*	.92	**.95**	.87
Binary, InfoGain	**.97**	.90	.85
Binary, GainRatio	**.97**	.90	.95
Frequency, no weights	.90	**.95**	.77
Frequency, *tf-idf*	.87	**.95**	.79
Frequency, InfoGain	**.92**	.85	.82
Frequncy, GainRatio	.92	.87	**.95**
SC metrics, no weights	**.79**	.77	.74
SC metrics, InfoGain	**.79**	.74	.72
SC metrics, GainRatio	**.79**	.74	.67

Table 7: Average classification accuracy using auto-cleaned transcripts. The highest classification accuracy for each feature set is indicated with boldface.

	Clean	Raw	Auto
PP rate	***	***	***
PP proportion	***	***	**
PP length	**		
VP rate		**	*
VP proportion	***	*	*
VP length	***		*

Table 8: Significance of the phrase-level features in each of the three data sets ($^{*}p < 0.05$, $^{**}p < 0.005$, $^{***}p < 0.0005$).

ficulty for the parser, which mis-labelled filled pauses and false starts in some cases. We also found that the insertion of filled pauses resulted in the creation of multiple features for a single underlying grammatical structure. The auto-cleaned transcripts appeared to avoid some of those problems, while still retaining information about many of the non-narrative speech productions that were removed from the clean transcripts.

Some of the features from the auto-cleaned transcripts appear to be associated with the discourse level of language, such as connectives and discourse markers. A researcher solely interested in studying the syntax of language might resist the inclusion of such features, and prefer to use only features from the human-annotated clean transcripts. However, we feel that such productions are part of the grammar of spoken language, and merit inclusion. From a practical standpoint, our findings are reassuring: data preparation that can be done automatically is much more feasible in many situations than human annotation.

6.2 Features

CFG production rules can offer a more detailed look at specific language impairments. We were able to observe a number of important characteristics of agrammatic language as reported in previous studies: fragmented speech with a higher incidence of solitary noun phrases, difficulty with determiners and possessives, reduced number of prepositional phrases and embedded clauses, and (in the raw transcripts), increased use of filled pauses and repair phrases. For this reason, we believe that they are more useful for the analysis of disordered or otherwise atypical language than traditional measures of syntactic complexity.

In some cases an in-depth analysis may not be required, and in such cases it may be tempting to simply use one of the more-general syntactic complexity measures. Nevertheless, even in our simple binary classification task, we found that using the more-specific features gave us a higher accuracy.

6.3 Future work

Because of the limited data, we consider these results to be preliminary. We hope to replicate this study as more data become available in the future. We also plan to examine the effect, if any, of the specific narrative task. Furthermore, we have shown that these methods are effective for the analysis of agrammatic aphasia, but there are other types of aphasia in which semantic, rather than syntactic, processing is the primary impairment. We would like to extend this work to find features which distinguish between different types of aphasia.

Although we included manually transcribed data in this study, these methods will be most useful if they are also effective on automatically recognized speech. Previous work on speech recognition for aphasic speech reported high error rates (Fraser et al., 2013a). Our finding that the auto-cleaned transcripts led to the highest classification accuracy is encouraging, but we will have to test the robustness to recognition errors and the dependence on sentence boundary annotations.

Acknowledgments

This research was supported by the Natural Sciences and Engineering Research Council of Canada and National Institutes of Health R01DC01948 and R01DC008552.

References

Harald Baayen, Hans Van Halteren, and Fiona Tweedie. 1996. Outside the cave of shadows: Using syntactic annotation to enhance authorship attribution. *Literary and Linguistic Computing*, 11(3):121–132.

Roelien Bastiaanse and Cynthia K. Thompson. 2012. *Perspectives on Agrammatism*. Psychology Press.

Julian Brooke and Graeme Hirst. 2012. Robust, lexicalized native language identification. In *Proceedings of the 24th International Conference on Computational Linguistics*, pages 391–408.

Jieun Chae and Ani Nenkova. 2009. Predicting the fluency of text with shallow structural features: case studies of machine translation and human-written text. In *Proceedings of the 12th Conference of the European Chapter of the Association for Computational Linguistics*, pages 139–147.

Eugene Charniak. 2000. A maximum-entropy-inspired parser. In *Proceedings of the 1st Conference of the North American Chapter of the Association for Computational Linguistics*, pages 132–139.

Franca Debole and Fabrizio Sebastiani. 2004. Supervised term weighting for automated text categorization. In *Text mining and its applications*, pages 81–97. Springer.

Kathleen Fraser, Frank Rudzicz, Naida Graham, and Elizabeth Rochon. 2013a. Automatic speech recognition in the diagnosis of primary progressive aphasia. In *Proceedings of the Fourth Workshop on Speech and Language Processing for Assistive Technologies*, pages 47–54.

Kathleen C. Fraser, Jed A. Meltzer, Naida L. Graham, Carol Leonard, Graeme Hirst, Sandra E. Black, and Elizabeth Rochon. 2013b. Automated classification of primary progressive aphasia subtypes from narrative speech transcripts. *Cortex*.

Harold Goodglass, Julie Ann Christiansen, and Roberta E. Gallagher. 1994. Syntatic constructions used by agrammatic speakers: Comparison with conduction aphasics and normals. *Neuropsychology*, 8(4):598.

Khairun-nisa Hassanali, Yang Liu, Aquiles Iglesias, Thamar Solorio, and Christine Dollaghan. 2013. Automatic generation of the index of productive syntax for child language transcripts. *Behavior research methods*, pages 1–9.

David I. Holmes and Sameer Singh. 1996. A stylometric analysis of conversational speech of aphasic patients. *Literary and Linguistic Computing*, 11(3):133–140.

Brian MacWhinney, Davida Fromm, Margaret Forbes, and Audrey Holland. 2011. Aphasiabank: Methods for studying discourse. *Aphasiology*, 25(11):1286–1307.

Ljiljana Progovac. 2006. *The Syntax of Nonsententials: Multidisciplinary Perspectives*, volume 93. John Benjamins.

Emily T. Prud'hommeaux, Brian Roark, Lois M. Black, and Jan van Santen. 2011. Classification of atypical language in autism. In *Proceedings of the 2nd Workshop on Cognitive Modeling and Computational Linguistics*, CMCL '11, pages 88–96.

Brian Roark, Margaret Mitchell, John-Paul Hosom, Kristy Hollingshead, and Jeffery Kaye. 2011. Spoken language derived measures for detecting mild cognitive impairment. *IEEE Transactions on Audio, Speech, and Language Processing*, 19(7):2081–2090.

Elizabeth Rochon, Eleanor M. Saffran, Rita Sloan Berndt, and Myrna F. Schwartz. 2000. Quantitative analysis of aphasic sentence production: Further development and new data. *Brain and Language*, 72(3):193–218.

Eleanor M. Saffran, Rita Sloan Berndt, and Myrna F. Schwartz. 1989. The quantitative analysis of agrammatic production: Procedure and data. *Brain and Language*, 37(3):440–479.

Kenji Sagae, Alon Lavie, and Brian MacWhinney. 2005. Automatic measurement of syntactic development in child language. In *Proceedings of the 43rd Annual Meeting on Association for Computational Linguistics*, pages 197–204.

Ben Swanson and Eugene Charniak. 2012. Native language detection with tree substitution grammars. In *Proceedings of the 50th Annual Meeting of the Association for Computational Linguistics*, pages 193–197.

Cynthia K. Thompson, Lewis P. Shapiro, Ligang Li, and Lee Schendel. 1995. Analysis of verbs and verb-argument structure: A method for quantification of aphasic language production. *Clinical Aphasiology*, 23:121–140.

Sze-Meng Jojo Wong and Mark Dras. 2010. Parser features for sentence grammaticality classification. In *Proceedings of the Australasian Language Technology Association Workshop*, pages 67–75.

Sze-Meng Jojo Wong and Mark Dras. 2011. Exploiting parse structures for native language identification. In *Proceedings of the Conference on Empirical Methods in Natural Language Processing*, pages 1600–1610.

Victor Yngve. 1960. A model and hypothesis for language structure. *Proceedings of the American Physical Society*, 104:444–466.

Author Index